Modern Fiction, Disability, and the Hearing Sciences

The relationship between critical disability studies and the hearing sciences is a dynamic one, and it's changing still, both as clinicians come to terms with the evolving health of deaf and hearing communities and as the 'social' and 'medical' understandings of disability continue to gain traction among different groups. What might a 'cultural' approach to these overlapping areas of study involve? And what could narrative prose in particular have to tell us that other sources haven't sensed?

At a time when visual media otherwise seem to have captured the imagination, *Modern Fiction, Disability, and the Hearing Sciences* makes the case for a wide range of literature. In doing so – through serials, short stories, circadian fiction, narrative history, morality tales, whodunits, the Bildungsroman, life-writing, the Great American Novel – the book reveals the diverse ways in which writers have plotted and voiced experiences of hearing, from the nineteenth century to the present day.

Edward Allen is Associate Professor in English at the University of Cambridge and Fellow of Christ's College.

Routledge Interdisciplinary Perspectives on Literature

Multilingual Currents in Literature, Translation and Culture
Edited by Rachael Gilmour and Tamar Steinitz

Rewriting the American Soul
Trauma, Neuroscience and the Contemporary Literary Imagination
Anna Thieman

Milton and the Early Modern Culture of Devotion
Milton and the Early Modern Culture of Devotion
Naya Tsentourou

No Dialect Please, You're a Poet
English Dialect in Poetry in the 20th and 21st Centuries
Edited by Claire Hélie, Elise Brault-Dreux and Emilie Loriaux

Mediterranean Slavery and World Literature
Captivity Genres from Cervantes to Rousseau
Edited by Mario Klarer

Literature with a White Helmet
The Textual-Corporeality of Being, Becoming, and Representing Refugees
Lava Asaad

Literary Fictions of the Contemporary Art System
Global Perspectives in Spanish and Portuguese
Carlos Garrido Castellano

Language, Style and Variation in Contemporary Indian English Texts
Esterino Adami

Modern Fiction, Disability, and the Hearing Sciences
Edited by Edward Allen

For more information about this series, please visit: www.routledge.com/Routledge-Interdisciplinary-Perspectives-on-Literature/book-series/RIPL

Modern Fiction, Disability, and the Hearing Sciences

Edited by Edward Allen

Routledge
Taylor & Francis Group
NEW YORK AND LONDON

First published 2025
by Routledge
605 Third Avenue, New York, NY 10158

and by Routledge
4 Park Square, Milton Park, Abingdon, Oxon, OX14 4RN

Routledge is an imprint of the Taylor & Francis Group, an informa business

Library of Congress Cataloging-in-Publication Data
Names: Allen, Edward, 1985- editor.
Title: Modern fiction, disability, and the hearing sciences / edited by Edward Allen.
Description: New York, NY : Routledge, 2024. | Series: Routledge interdisciplinary perspectives on literature | Includes bibliographical references and index.
Identifiers: LCCN 2024007388 | ISBN 9780367261306 (hardback) | ISBN 9781032799476 (paperback) | ISBN 9780429291616 (ebook)
Subjects: LCSH: English fiction—History and criticism. | American fiction—History and criticism. | Disabilities in literature. | Hearing disorders in literature. | LCGFT: Literary criticism. | Essays.
Classification: LCC PR830.D558 M63 2024 | DDC 813.009/35272—dc23/eng/20240317
LC record available at https://lccn.loc.gov/2024007388

ISBN: 978-0-367-26130-6 (hbk)
ISBN: 978-1-032-79947-6 (pbk)
ISBN: 978-0-429-29161-6 (ebk)

DOI: 10.4324/9780429291616

Typeset in Sabon
by Apex CoVantage, LLC

Contents

Contributors

Edward Allen, *Christ's College, University of Cambridge*

William Allen, *University of Glasgow*

Catherine Charlwood, *University of Exeter*

A. Elisabeth Reichel, *Osnabrück University*

Rachel Farebrother, *Swansea University*

Andrew Gaedtke, *University of Illinois, Urbana-Champaign*

Ben Holmes, *Birmingham and Solihull Mental Health NHS Foundation Trust*

Beryl Pong, *Trinity College, University of Cambridge*

Anna Snaith, *King's College London*

Justin Tackett, *North Carolina State University*

Jaipreet Virdi, *University of Delaware*

Clare Walker Gore, *Lucy Cavendish College, University of Cambridge*

Acknowledgements

Thanks, first, to those who attended 'Ear Pieces: Listening, Diagnosing, Writing' in late 2016. This interdisciplinary conference was funded by a medical humanities grant, courtesy of the Wellcome Trust, and hosted by the Faculty of English in Cambridge. Among those I'd like to pick out for particular thanks are Carolyn Abbate, Steven Connor, and Mara Mills; and Laura Davies, Fiona Green, Marica Lopez Diaz, Esther Osorio Whewell, Jen Pollard, David Trotter, and Flora Willson.

In the spirit of that conference, this volume of essays has evolved in surprising ways and been nourished by advice from many quarters. In this regard, I'd like to thank Semele Assinder, Mel Bach, Rebecca Fleming, Graeme Gooday, Olivia Marshall, Sarah Meer, and Sophie Read. Huge thanks are due, too, to Jennifer Abbott and Anita Bhatt at Routledge for seeing the book through – so patiently – and to the following institutions for permission to reproduce material: Curtis Brown, London, on behalf of the Trustees of the Mass Observation Archive © The Trustees of the Mass Observation Archive; the National Archives; *The Musical Times*; and Punch Limited.

Thanks, finally and especially, to the late David Baguley (1961–2022). David – Dave – was an early, generous supporter of this project. Our paths crossed shortly before he moved to Nottingham from Cambridge to take up the role of Professor of Hearing Sciences. But he came back to Cambridge for 'Ear Pieces' – as keen to muck in as to provide expert insight – and he was nothing but supportive as my thoughts turned from that conference to this book, to which he agreed to write an afterword. Though it is poorer for the absence of that piece, I hope this book does justice to the quality and scope of Dave's aspiration with regard to multidisciplinary research, which was simply, cheerfully, to get the 'two cultures' talking.

Introduction
Placing Quietness

Edward Allen

DAY 89. With a cymbal roll and falling pitch-bend, the black-and-white intertitle gives way to a burst of muted colour. Red, amber, green: the unmistakable sight – a set of traffic lights – lies powerless in the dust. By inviting the eye to roam the depths of each lens, the film's opening close-up reveals a concentric, stippled effect, a bejewelled pattern ordinarily effaced by the lights' automated operation. 'We begin to confront the thingness of objects when they stop working for us'.[1] The 'when' we can guess, thanks to that preliminary intertitle, but the reason for the 'stop', and the whereabouts of the 'us' for whom the object-turned-thing is no longer working, are harder to fathom. The defunct signal head makes for a lonesome, textured image – blurred at the edges, broken here and there by seed heads and encroaching weeds – and it makes for a textured soundscape, too. Crisp wisps of grass crackle in the wind around it, the distant tap-tap and creak of rusted metal – a hinge? a sign? a flagpole's halyard? – hint at a world left to its own devices. Piecing together what's happened seems the only reasonable thing to attempt as a viewer, a reparative thing, even, given the intimations of recent apocalypse. Yet 'it might be more true to say that narrative has in fact been suspended', as David Trotter observes, scratching at the mode he calls 'rust-belt Gothic'.[2] For this is a mode of representation in which storytelling has stalled or suffered evacuation, and 'description' – or the failure 'to sort the significant from the insignificant' – has taken its place.[3] Nothing but a static camera-eye survives to drink in the detail of this wasteland. Nothing but a camera-eye and a microphone, that is, for this is *A Quiet Place*, not a silent one.

Since you've chosen to open a book with 'hearing' in its title, you'll probably be unsurprised to have your attention drawn to this distinction. But is there more to *A Quiet Place* than meets the eye, or the ear? Moving back to the film, and moving from the outside in, we're prompted to survey the ruined interior of Larkin's Market. It's a dead space, and a deadening acoustic, but for the sudden patter of tiny feet. Left to right, a little figure dashes barefoot, and what might have been a jump-scare dissipates

DOI: 10.4324/9780429291616-1

Figure 0.1 Opening shot, *A Quiet Place* (2018)

into a game of hide-and-seek: another figure tiptoes along the grimy floor, glancing right then left, while the first scurries away into an intersecting aisle. It could be a video game. As the figures appear to multiply – a boy crouches against some shelving, a woman searches a medicine cabinet – so do our questions about the scene's audio design. The wind in the street supplies a gentle white noise, overlayed with the mystery shoppers' steps and breathing; but this incidental wash of sound comes and goes, subtly so, in keeping with one of the roaming figures' sense perceptions. She of the tiptoeing tactic advances towards the camera, then past it, and as she does so a first glimpse of her cochlear implant effects a switch in the composite soundtrack's frequencies, from white noise to heartbeat – seven unmistakable pulses – and then back again as the camera moves on.

It's a neat gimmick, and it won't be the last time the implant comes into its own. For now, it takes us into the body of its owner, Regan Abbott, into her biorhythms, into an internally focalised position from which the film's descriptive exposition begins to make sense, begins to *mean* something. The next time we see and hear Regan up close, she's knelt in an adjacent aisle with her brother, Beau, watching as he puts the finishing touches to a chalk drawing on the floor. He signs to her by way of explanation:

Beau: Rocket.
Regan: Very good job!
Beau: That's how . . . we'll get away.[4]

Figure 0.2 Regan, Beau, and the chalk drawing, *A Quiet Place*

Though it takes just a second or two in real time for this child's play to play out, the siblings' interaction should give us pause. Beau turns his hand to the signified (rocket) in two ways – sketch and sign – and his translation is instantly overlayed by the flash of a subtitle. Sketch to sign, sign to subtitle, this triangulation signifies something other than the thing itself, for it exposes a discrepancy between one language system and another – between the arbitrariness of 'rocket' (the word bears no resemblance to the concept) and the iconic properties of its equivalent in American Sign Language (which involves forming an 'R' with your dominant hand and launching it, rocket-like, from the base of your non-dominant) – but to what purpose? The answer has to do with Beau's hushed attempt to make the stuff of dreams legible, and with his doing so in Regan's presence. Resonating as no spoken or written word could, from a crisscrossing of the index and middle fingers ('R') to its sharp, rising trajectory, Beau's 'rocket' takes shape in a qualitatively different way to the formation of a vocal or scripted utterance, and we apprehend that difference because we've been tuned in to Regan. To inhabit her personal acoustic, to detect and acclimatise to those lower, scarcely audible frequencies, even to feel you might have got inside her head – this trick of audio-visual engineering sensitises you anew to the intricacies of multimodal perception. Under these hyper-focalised conditions, Beau's message seems more pronounced, his signing more significant, his rocket somehow – yes – more rocket-like.

Only *seems*, however. As surely as the boy seems to spell things out – in fact, the subtitles do that, and they do that for the viewer, not for him or Regan – the scene is curated in such a way as to feel consequential even as the meaning generated between the siblings is deferred. Beau suddenly scarpers off to find a concrete referent – a toy space shuttle perched on a nearby shelf – and we remain with Regan, mutually absorbed by the aborted interaction. It's too early to attribute symbolic value to the imagined escape craft. Yet something of that order does hang suggestively in the air – the memory of a finger-rocket arcing quietly into space between a brother and a sister – counterpoised by the sheer gravity of Regan's look. Already, Beau's 'very good job' has the aura of something else. The hint of a 'deafened moment', perhaps, as Lennard Davis once put it, but a warning sign, too.[5]

* * * *

A Quiet Place feels the right place to begin a book about disability and the hearing sciences, for two quite different reasons. For the scholar of deaf studies, the film's a dream. It's not only that it seems to have it all – deafness and hyperacusis, lip-reading and ASL – nor that its acquisitive approach to the hearing sciences screams a sense of urgency ('IT'S SOUND!') not often projected in movie theatres. As its opening sequence shows, *A Quiet Place* is worth watching closely because it feels its way into the multimodal texture of deaf life. It feels its way heuristically, that's to say, without fetishising or seeking to exploit that way of life, as another film might, for cheap thrills or prolonged voyeuristic pleasure. As it transpires, *A Quiet Place*'s genius hinges on the premise that a deaf life is now the *only* way to live. Earth has been colonised by an alien species with a more-than-usually-acute sense of hearing (though who knows what's normal for aliens; maybe we're the unusual ones). Since these invaders depend on echolocation to chase down all those clumsy, fidgety humans, the ability to sign has become an 'evolutionary advantage' for the Abbott family, as commentators have noticed, and it's because the whole family can do it – not just Regan – that the film rapidly comes to seem greater than the sum of its parts: not an insight into so many hearing lives lost or subdued, but a chronicle of human culture properly adapting to a new (deafened) normal.[6] For this and other things, the critics loved it. '[T]he whole saga is shaped around its sound design', Anthony Lane observed, having commended the film's 'splendid premise'; while T. Frohock was glad to find 'a very intense human drama' at the heart of what might otherwise have been 'a shoot-em-up, run-around-and-get-slashed horror movie'.[7] Peter Bradshaw went the whole five stars. 'In its simplicity and punch, this is a film that feels as if it could have been made decades ago, in the classic age of *Planet of the Apes* or *The Omega Man*'.[8] Except, of course, for the high-concept audio, which seemed to Bradshaw nothing short of headline news: 'silence never sounded so terrifying'.

Even as he begins to sound his years – conjuring precedents and classics, invoking the history of the 'back-to-basics thriller' – Bradshaw strikes a contemporary note in suspecting *A Quiet Place* of coaxing us into *un*precedented territory. Yet it's not so much the sage reviewer's voice we hear in that soundbite – *silence never sounded so terrifying* – as the rumble of a movie trailer voiceover: ominous, bassy, grounded in a present perfect tense that makes the essence of the claim (owing to the missing auxiliary) that bit harder to historicise. But historicise it we must. *Has* silence never sounded so terrifying? Though primarily rhetorical, the force of such a question has the potential to disrupt our attention, our expectations, at what feels like a crucial juncture in the history and representation of the senses. To one way of thinking, *A Quiet Place* is evidently that rare thing – a thoughtful and thought-provoking blockbuster – and to this degree we must assume it would have puzzled those mass-media sceptics at mid-century, hell-bent on demonising the 'big film' (*Großfilms*): this is not, after all, a film that 'seeks strictly to reproduce the world of everyday perception'; in point of fact, it takes great pains to achieve the opposite – to scramble and disorientate – and perhaps, in doing so, shows signs of that *Unbotmäßigkeit*, or 'unruliness', for which Adorno and Horkheimer once appeared so misty-eyed.[9] Yet members of the Frankfurt School were not wrong to lament the schematic tendencies of the culture industry, or to suppose that 'the sound film' had become – and would remain, as indeed it has – its 'most characteristic' product.[10] Whatever its merits and whatever its reviewers' views ('silence never sounded so terrifying'), *A Quiet Place* must also be judged to have contributed to a wave in cinema no less schematic than those of the mid-century stunt film or cartoon. '[A] recent rash of films offers a tourist's introduction to the Deaf world', Lennard Davis reported in September 2021. 'In nearly all these films', – *A Quiet Place* (2018), *Sound of Metal* (2019), *CODA* (2021), and the inevitable sequel, *A Quiet Place Part II* (2021) – 'being Deaf is treated as a problem that the film has to fix in some way'.[11]

So here's the second, polemical reason for commencing with *A Quiet Place*. It's not that the film hasn't important things to say about disability and the hearing sciences; it's because it appears to do what it does so adeptly – ably, to Bradshaw's mind, and ableistically, to Davis's – that one might suspect it not only of convincing viewers that deafness is indeed 'the next big thing', but also of absorbing us in such a way as to relegate other kinds of material and other modes of imagining to the periphery.[12] Why roam elsewhere when there's *A Quiet Place* to explore, a diegetic soundscape so immersive as to render soundscapes of the dim and deafened past null and void? The answer posed by this collection of essays is by definition implicit: nothing does silence like modern fiction. Or, to adapt the popular lyric, nothing does the *sound* of silence like modern fiction. As one of the contributors to this collection has stated before, and just as beautifully as Paul Simon:

> the literary text [is] a site of vocalisation, rhythmics and dissonance as well as an archive of soundscapes and modes of listening. Putting your ear to a book – touching skin to paper – yields much more than resounding silence. The sonic takes us to literature's place in the world – how it is heard, read, declaimed – as well as its representation of multisensory experience.[13]

A text can seem a quiet place, but this must be considered a strange fiction, and two words here – 'site' and 'archive' – indicate right away that the sorts of resonance achieved in literary form require a reciprocal act of imagination that is different in kind from the make-believing peculiar to a booming multiplex. To think of a text as a site might be to think of it set apart yet always already in situ, not so much a closed acoustic as the co-ordination and emplotment of voices in relation to a particular environment. However mobile or vulnerable to displacement it may then prove to be – its subsequent circulation among readers is never beside the point, nor its remediation – a text conceived in these site-specific terms is likely to reward our circling back to it, to reward a mode of reading (however 'silent') that is newly attentive to the interface between it and the evolving practices which constitute its historical moment. Although any text, in other words, could be thought to possess a timeless, internal dynamic – an arrangement of voices, a bespoke sound system of sorts – this supposed dynamic is always contingent, though not always transparently so, on the lived experiences of voicing and listening in time. It matters to *The Years* (1937), to its temporal and sonic textures, that Virginia Woolf found herself conflicted in its writing, caught between sounds as well as wars, sirens and amplified demagogues. Just so for Martin Amis and *Money* (1984): his is a novel in which hearing things – an idiom so often uttered as to have become hollow – is given a name and made to sound at once new and painfully *nouveau-riche*. Inasmuch as tinnitus is perceived in Amis's book to signal a level of disquiet appropriate to the age of deregulation, one word alone would seem to encapsulate the interaction between hearing complaint and historical malaise for which fiction supplies the binding site: the word – as Elisabeth Reichel suggests – is *symptomatic*.

That this word has since come to typify a mode of reading whose currency was on the rise when *Money* appeared in 1984 – a mode of interpretation commonly understood to privilege the unsaid or unsounded, the barely murmured, the absent presence ('like a word on the tip of your tongue') – may come as a salutary reminder that literary criticism is well used by now to detecting the inaudible, or to the fantasy of doing so.[14] What it might mean to divulge 'silences, gaps', or in Jameson's terms to 'reaudition' marginalised voices of the past, will certainly be a matter of concern for any book invested, as this one is, in the representation of disability.[15] In the end, however, this is not so much a book about 'symptomatic' reading as it is a collective effort to recuperate the clinical resonances of that word. To treat it once more as a point of convergence, as a category shared and refined between disciplines, is to follow the lead of David Baguley:

> When a clinician or scientist specializes in a particular condition or symptom, he or she can become vigilant to mentions of that symptom in literature and music. The same can be true of patients, who may be hyper-alert to and troubled by mentions of their symptom in a negative or catastrophized way.[16]

If this captures a mood of heightened attention, it does so with a doctor's caution, mindful of the risks involved in trusting to a false equivalence between close reading and clinical diagnosis. Yet reading the clinician closely reveals a no less prudent resolve to countermand the kind of vigilance that might foreclose or limit the scope of the medical humanities in its most imaginative humour. Baguley has read his Amis, and his Bible and Hardy and Rowling, so he can state with confidence that reading widely 'reminds us as clinicians of the massive heterogeneity of responses to tinnitus in the population, and not to oversimplify the situation'.[17]

It is in Baguley's spirit that we set out in this book to chronicle a longer, more variegated history of disability and the hearing sciences than the type we find reduced to a flashpoint in *A Quiet Place*. If it's right to think of a text or texts constellating to form an 'archive' – a repository, in Anna Snaith's phrase, of 'soundscapes and modes of listening' – it's worth noting that the archive assembled here is indicative rather than exhaustive, creative rather than schematic, the better to suggest not only possibilities for future work but some criteria, too, for judging the best way to model a 'cultural' understanding of disability. With a view to loosening the binary definition of disability that seems to have been baked into critical debate – 'medical pathology' versus 'social discrimination' – such an approach depends to a large degree on stories, and on recognising the telling of stories for what it is: not a decorative or trivial pursuit but a human means of meaning-making through which any society might come to think of itself and its definitions, lived and legislated, as *cultured*.[18] Rather than prompting us to either avail ourselves or dispose entirely of the terms that have historically come between the medical and social models – impairment and disability – storytelling compels us never to take the semantic contours of the debate as read. It reminds us, too, insofar as terminology always trips from some hand or tongue, to look for the thing that sometimes appears to have got lost in culture wars: not the categories in identity politics, but the characters.

If 'literary writing', as Alice Hall remarks, 'serves a pedagogic function in the sense that it not only documents but also shapes attitudes towards disability', it must also be said that literary writing – or the writing on show here, at any rate – has less to do with teaching us what to think and more to do with schooling us in the pleasure of thinking differently, *through* different characters, not in spite but precisely because of the stories we otherwise like to tell ourselves.[19] So many varieties of fiction show deafness to be exactly that – a fiction, a matter of perspective – in the sense that 'it' has rarely been reducible to one thing; as Mara Mills argues, '"deafnesses"' would be a more nuanced way of accounting for a range of hearing loss, and of acknowledging the several dimensions in which a deaf person's experience of sound manifests, since any such experience is bound to be – in Rebecca Sanchez's words – 'multisensory and multimodal, prosthetic, and interdependent'.[20] It may, of course, be no less of a fiction to think the plural form of deafness the last word in this conversation: but for the analytical granularity it promises to effect, even 'deafnesses' may prove to be too essentialising if it's allowed to obscure or flatten out the differences of 'hearing' differently in a continuum that is, after all, of interest as much to brain science as to otological medicine. Additional terms of reference, and not just those centred on the vestibular apparatus – subaudition, for instance, and auditory hallucination, schizophonia and articulate silence – may bring us that bit closer to knowing what the thing feels like, which sometimes isn't (or isn't just) a malfunction of the inner ear, but a blend of neuro-auditory phenomena that can make it hard to locate the source of what's going on. For this reason alone it seems right to have the first and last pieces in this volume turn deafness on its head, the first by contemplating the extension of apparently healthy ears to unhealthy bodies, the last by reflecting on what it means to 'really' listen – even if it means hearing too much – in cases of dementia.

As well as attending to different kinds of received storytelling – serials, short stories, circadian fiction, narrative history, morality tales, whodunits, the *Bildungsroman*, life-writing, the Great American Novel – this book is also interested in storytelling now. Interleaving chapters written in a more or less literary-critical mode is a trio of tales told from the front line of health and social care. Each of these 'Ear Pieces' is anecdotal, more personal than an essay, though true to the spirit of that genre in trying or trialing a different form of commentary; no less than the stories considered elsewhere in the collection – but with an added edge, given the intimate relation in each case of narrator to subject – they bring our primary concerns into focus even as they counterpoint the collection's critical rhythms. The tense conceptual interplay in disability studies of deficit and asset. The urge to witness, to plot and voice, a growing body of knowledge that not only requires communication but has the idea of it at its heart. Even the idea that stories could have a therapeutic function.[21] This book is about all these things, though in the end our collective brief could be expressed by turning the words of one of the contributors into a question. What does it mean 'to be prepared to find out what you don't know'? The purpose of this book is not so much to supply an answer to that as to explore, through the writing and reading of fiction, the untold complexity of those monosyllables when they're brought to bear on the hearing sciences. Given an unfamiliar soundscape, what *does* it mean to 'find' things out? What *could* it mean to 'know' something you didn't know before when knowing no longer means hearing, or when knowing can only mean hearing more or less than the unspoken norm? These questions have a history, as we're going to find out now.

Notes

1 Bill Brown, 'Thing Theory', *Critical Inquiry*, 28.1 (2001), 1–22 (p. 4).

2 David Trotter, 'Rust-Belt Gothic: The New Technology Narrative', *The Lit Platform* (2012). https://theliteraryplatform.com/news/2012/12/rust-belt-gothic-the-new-technology-narrative/ [accessed 29 October 2021]

3 Ibid.

4 *A Quiet Place* (Paramount Pictures, 2018), screenplay by Bryan Woods, Scott Beck and John Krasinski, directed by John Krasinski.

5 Lennard J. Davis, *Enforcing Normalcy: Disability, Deafness, and the Body* (London and New York: Verso, 1995), pp. 100–1.

6 Simon Bacon, *Vampires from Another World: The Cinematic Progeny of H. G. Wells'* The War of the Worlds *and Bram Stoker's* Dracula (Jefferson, NC: McFarland, 2021), p. 168.

7 Anthony Lane, 'Heard Instinct', *The New Yorker* (16 April 2018); T. Frohock, '*A Quiet Place*: A Review from the Deaf Perspective' (23 April 2018): <www.tfrohock.com/blog/2018/4/23/a-quiet-place-a-review-from-the-deaf-perspective> [accessed 29 October 2021].

8 Peter Bradshaw, 'Silence Never Sounded So Terrifying [*A Quiet Place* Review]', *The Guardian* (5 April 2018): <www.theguardian.com/film/2018/apr/05/a-quiet-place-review-john-krasinski-emily-blunt> [accessed 29 October 2021].

9 Max Horkheimer and Theodor W. Adorno, 'Kulturindustrie', in *Dialektik der Aufklärung: Philosophische Fragmente* (Frankfurt am Main: Fischer Taschenbuch Verlag, 2006 [1969], pp. 133–6); 'The Culture Industry', in *Dialectic of Enlightenment: Philosophical Fragments*, ed. Gunzelin Schmid Noerr, trans. Edmund Jephcott (Stanford: Stanford University Press, 2002), pp. 99–101.

10 Adorno and Horkheimer, 'The Culture Industry', p. 100.

11 Lennard J. Davis, 'Screening Deafness', *Los Angeles Review of Books* (19 September 2021). https://lareviewofbooks.org/article/screening-deafness [accessed 29 October 2021].
12 Lennard J. Davis, 'Quiet Places', *Los Angeles Review of Books* (20 April 2018). https://lareviewofbooks.org/article/quiet-places/ [accessed 29 October 2021].
13 Anna Snaith, 'Introduction', in *Sound and Literature*, ed. Anna Snaith (Cambridge: Cambridge University Press, 2020), pp. 1–33 (2).
14 Fredric Jameson, *The Political Unconscious: Narrative as a Socially Symbolic Act* (Abingdon: Routledge Classics, 2002 [1981]), p. 121.
15 Stephen Best and Sharon Marcus, 'Surface Reading: An Introduction', *Representations*, 108.1 (2009), 1–21 (p. 6); Jameson, *The Political Unconscious*, p. 71.
16 David M. Baguley, 'Tinnitus and Hyperacusis in Literature, Film, and Music', in *Tinnitus: Clinical and Research Perspectives*, ed. David M. Baguley and Marc Fagelson (San Diego: Plural, 2016), pp. 1–11 (1).
17 Ibid., p. 9.
18 For a helpful overview of the 'medical' and 'social' models, and for the significance of a third way, see Andrzej Twardowski, 'Cultural Model of Disability – Origins, Assumptions, Advantages', *Kultura i Edukacja*, 2.136 (2022), 48–61 (p. 52).
19 Alice Hall, *Literature and Disability* (London and New York: Routledge, 2016), p. 4.
20 Mara Mills, 'Deafness', in *Keywords in Sound*, ed. David Novak and Matt Sakakeeny (Durham: Duke University Press, 2015), pp. 45–54 (45); Rebecca Sanchez, 'Deafness and Sound', in *Sound and Literature*, pp. 272–86 (273).
21 Baguley, 'Tinnitus and Hyperacusis', p. 9.

1 Stethoscape
Auscultation in British Fiction

Justin Tackett

The stethoscope was invented by French physician René Laennec when he rolled 'a quire of paper into a sort of cylinder and applied one end of it to the region of the heart and the other to [his] ear'.[1] In 1819, he published *De l'Auscultation Médiate* – the first treatise on 'mediate auscultation', the practice of listening to the body's interior using a stethoscope. Laennec's technique was 'mediate' because it was *mediated* by the stethoscope; it improved upon the older practice of 'immediate auscultation', listening to the body's interior by applying the ear directly to the body. It also surpassed 'percussion', the practice of striking the body with fingers or a pleximeter and listening to the sound it made, which was developed by Leopold Auenbrugger in the 1760s. Laennec's mediate auscultation was successful because it funneled and amplified the body's sounds better than other techniques, and for this reason it soon became known simply as 'auscultation'. In contradistinction to predecessors such as the ear trumpet, the stethoscope should be thought of as an *active* tool. Auscultation does not just receive sounds that anyone can hear but packages those sounds as hearable for specialized listening, what Jonathan Sterne has called 'audile technique'.[2] The mere presence of a stethoscope is solidified sound: it always implies there is something to be heard.

In addition to its French development, auscultation has a unique history in British science and literature as well as a significant role in realistic fiction from the 1830s onward. From the beginning, auscultation disclosed individuals' interiors – by figurative extension, their interiorities – and it was a social tool that drew boundaries between professionals and laypeople, men and women, healthy and diseased, as many scholars have shown. As much as doctors were concerned with demonstrating their literary knowledge – what Michael Brown has called 'medico-gentility' – authors, another professionalizing occupation, could pride themselves on demonstrating medical knowledge too.[3] For such reasons, auscultation was ripe for appearance in realistic fiction across the nineteenth century. But I argue in this chapter that auscultation in fiction is not simply a feature of Roland

DOI: 10.4324/9780429291616-2

Barthes's 'reality effect', nor is it a tool that merely encourages us to read beneath the surface.[4] Building upon scholarship by Lawrence Rothfield, John Picker, Richard Menke, and others, I argue that auscultation additionally exhibits a particular mechanism of realism, akin to Foucault's *visible invisible*.[5] Auscultation contributes to realism precisely because it makes conspicuous that which cannot be known. I argue that the stethoscope particularly evinces a realism that thrives on prohibiting access to the real and therefore encourages recognition of a different way of listening to the period's fiction, what I formulate as 'subaudition' in the second half of this chapter.

In attempting to approach a summative (but by no means exhaustive) claim, I have surveyed roughly ninety works of fiction that portray or allude to auscultation, mostly novels but also novellas, short story collections, and even cartoons, in addition to medical manuals, lectures, articles, tables, and diagrams. For this reason, what follows consists less of sustained close readings and more of extensive samplings of fiction across the century, much in the spirit of a 'culture of the excerpt'.[6] First, I provide a brief history of the peculiarities of auscultation in Britain pertaining to the English language. Next, I explore doctors' and authors' preoccupations with professionalization in medicine. In the remainder of the chapter, I examine how auscultation participates in 'subaudition', a Victorian kind of semi-understanding that bolsters realism.

Before proceeding, I wish to clarify what I mean by 'realism' in this chapter. Which features constitute the genre 'realism' and which writing counts as 'realist' has long been the subject of debate.[7] Indeed, the categorization of several of the authors and works that follow has often been at the centre of that debate. But I am less interested in fretting the boundaries of a genre than I am in identifying how scenes of auscultation operate in British fiction. My procedure has been to start with such scenes, analyze as many as possible, and ask what patterns emerge; I have not, in other words, started with 'realist' works to the exclusion of other genres, then found auscultation scenes within them. Appeals to the real – desiderata of what Ann Elizabeth Gaylin calls the 'epistemophilic quest' in her study of eavesdropping – emerged as one such pattern in the scenes I encountered, though they can (and do) appear in fiction that contains elements of unapologetic fantasy as well.[8] If this selection of fiction 'unhitch[es] the Victorian novel from realism' while hitching others, that is somewhat beside the point.[9] For these reasons, I opt for 'realistic' instead of 'realist' wherever possible; 'realistic' has the additional advantage that it entered English around the 1820s, when auscultation came to Britain.[10] For stylistic reasons, however, I have also defaulted to using 'realism' to refer to techniques evincing reality. In what follows, I show how auscultation is a conspicuous example of Audrey Jaffe's '*realistic fantasy*', an emblem that positions readers and characters 'in a seductive and yet impossible-to-fulfill relation to a fantasized real', in this case the 'reality' of sound that is impossible to hear.[11] I would add that the history of the stethoscope (indeed, the perspective of modern medicine in general) was complicit in this dynamic, reaffirming that 'both the scientist and the novelist are engaged in what is typically called a novelistic [pursuit], and that science too is a way of reading'.[12]

Part 1: A Brief History of Auscultation in English

The manner in which British doctors imported auscultation from France after Laennec published his treatise was entangled with preoccupations about Britain's own medical contributions and the English language. Anglicizing auscultation was not just a matter of disseminating knowledge but of wresting it from its continental origins by providing alternative British histories, reforms, and, in one particular case, a misnomer that became a novel reality. In each instance, auscultation was shaped and reshaped by British linguistic ideologies that governed and were governed by fiction.

Several British doctors were key to importing auscultation into English. James Clark was the first to describe auscultation and the stethoscope, which he witnessed at Laennec's hospital in Paris, in *Medical Notes on Climate, Disease, Hospitals, and Medical Schools* (1820). (I will point out an important peculiarity of Clark's book later in this section.) The following year, John Forbes translated Laennec's work into English as *A Treatise on the Diseases of the Chest* (1821); Forbes's translation was highly influential and went through several expanded editions. Lastly, Peter Mere Latham published *Lectures on Subjects Connected with Clinical Medicine* (1836), the first original English composition on auscultation. Thus, the close of the Regency period saw the circulation and gradual adoption of auscultation among doctors. By the late 1830s, though, only a few novels had mentioned auscultation and the stethoscope.

The Victorian period was marked by a much wider acquaintance with auscultation, not least because the evolution of the stethoscope and the basic principles of auscultation were highly publicized. Whereas the earliest stethoscopes were simple foot-long tubes of wood, ivory, and brass, the stethoscope became more malleable and flexible when Britain's imperial expansion made gutta percha and rubber available. The stethoscope also sprouted its now-distinctive double earpieces at the Great Exhibition of 1851, where Irish physician Arthur Leared debuted the binaural stethoscope. As public curiosity grew, physicists such as John Tyndall began giving public transatlantic lectures in the 1860s and '70s on acoustics, which included auscultation. During this time, auscultation spread among Britain's echelons as women took medical degrees and throughout 'Greater

Britain' as peoples from North America, Africa, and Asia came to Britain to study medicine, frequently returning to their home nations wielding stethoscopes. Treatises advocating auscultation, such as an 1874 book on tropical diseases by Africanus Horton, the first Nigerian (of at least eight in the nineteenth century) to earn a British medical degree, might be focused on European bodies; written in Western Africa; dedicated to a physician in India; and published in London.[13] Such was auscultation's cosmopolitanism.

Because of its distinctive shape and widespread use, the stethoscope became the preeminent symbol (rivaling the asklepian and red lamp) of the medical profession and was universally recognized by doctors and laypeople by the end of the century. Auscultation was widely seen as demarcating the end of one period and the beginning of another, a symbol of 'the conflict between the old and the new systems of medical science', according to one early novel.[14] Or, as George Eliot reminds us in *Middlemarch* (1871–72), writing from the Victorian era about the previous age, auscultation 'had not become a matter of course in practice at that time', having been invented by 'Laennec [. . .] not so very many years ago'.[15] Thus, while auscultation was a Regency invention, it was a Victorian phenomenon.

Perhaps anxious to assert Britain's intellectual prowess as global power dynamics shifted across the channel from Napoleonic France, British scientists attempted to claim auscultation for their nation. Both Forbes and Tyndall, for example, claimed that Robert Hooke had anticipated Laennec's invention in *Posthumous Works* (1705), where Hooke speculates that 'Otocousticons' may aid the ear, just as telescopes and microscopes aid the eye. Through augmented hearing,

> it may be possible to discover the Motions of the Internal Parts of Bodies, whether Animal, Vegetable, or Mineral, by the sound they make [. . .]. [O]ne may discover the Works perform'd in the severals [*sic*] Offices and Shops of a Man's Body, and thereby discover what Instrument or Engine is out of order, what Works are going on at several Times, and lie still at others, and the like. [. . .] I have been able to hear very plainly the beating of a Man's Heart, and 'tis common to hear the Motion of Wind to and fro in the Guts, and other small Vessels, the stopping of the Lungs is easily discover'd by the Wheesing, the Stopping of the Head, by the humming and whistling Noises, the sliping [*sic*] to and fro of the Joynts in many cases, by crackling, and the like.[16]

Forbes and Tyndall argue that Hooke prophesied auscultation in this passage (even as they give due reverence to Laennec's accomplishments) and so partially retrieve the technique as British in concept. The passage might be significant to British fiction as well, since Hooke speculates that 'in Plants and Vegetables one might discover by the Noise the Pumps for raising the Juice, the Valves for stopping it, and the rushing of it out of one Passage into another, and the like'. While Picker has noted the impact Laennec, Tyndall, and Hermann von Helmholtz likely had on George Eliot, for example, Hooke's speculations here about plants and animals provide another possible genealogy for 'hearing the grass grow and the squirrel's heart beat', which roar on the other side of silence.[17]

Whatever the case, it was a trope in British fiction for old-fashioned doctors to regard auscultation with suspicion because it was (too) French. A doctor who auscultated might be identified in a British novel as a 'dapper little Frenchman', for example, and Arthur Conan Doyle characterized one of his curmudgeonly fictional doctors as being known to say 'vain things about Laennec, and to refer to the stethoscope as "a newfangled French toy"' seventy years after it had been invented.[18] In one curious instance, Laennec himself appears in Rudyard Kipling's 'Marklake Witches' (1910). Set during the Napoleonic wars, the tale portrays Laennec as a French prisoner in an English village, where the local British doctor declares him 'worse than [an] impostor' and denounces the stethoscope ('trumpet') as 'some papistical contrivance'. Laennec's British host defends the French physician not by endorsing auscultation but by declaring (accurately) that Laennec is 'a Breton, which is nearly next door to being a true Briton'.[19] Just as Forbes, Tyndall, and (perhaps) Eliot sought to anglicize auscultation by invoking Hooke, so the characters in 'Marklake Witches' seek to make Laennec as British as possible.

But beyond the nationalist and imperialist histories that British doctors and writers attempted to apply to auscultation, a deeper impulse shaping its evolution was the project of bringing the technique into the English language. Parisian doctors constructed elaborate taxonomies of the sounds they were hearing, aiming for a precision they hoped would uniquely match sound to disease. In the mid-1830s, for example, Adam Raciborski – a Polish physician practising in Paris – wrote manuals on auscultation, and an 1839 edition in English came with an unfoldable folio-sized 'synoptic table' that attempted to organize these sounds. The table includes: 'Tic tac, or normal sound of the heart'; 'Dry or grave sonorous rattle'; 'bass chord or string'; 'Ronchus'; 'Œgophony. Goat-like voice'; 'Punch'; 'blowing through a comb'; 'Stammering'; 'Saw and rasp sounds'; 'Grazing'; 'Friction'; 'New leather'; 'Scraping'; 'Bellows'; 'Metallic tinkling'; 'metallic clashing'; 'Continuous bellows'; 'Bruit de diable'; 'modulated hissing'; 'parchment when rumpled'; 'rattle'; 'cooing of turtle doves and the cries of different birds'; 'rubbing stiff hair between the fingers'; 'English taffeta'; 'drop of water falling into a decanter already containing a certain quantity of fluid'; 'gentle and quick friction of one finger against the other, so as to produce a kind of fillip'; 'whizzing sound'; 'cracked pot when struck';

'buzzing of a fly in a large vase'.[20] Such auscultative sounds were legion and multiplying.

British doctors balked. Participating in the nineteenth-century positivism that would engender twentieth-century philosophy of language (and recalling the conviction of Englishman William of Ockham), they were cautious not to multiply entities beyond necessity and aimed to tame the French sonic menagerie. It was the duty not just of the scientist but of the *British* scientist to reduce and generalize. Forbes cautioned against empty names and fabricating internal conditions that are not real:

> [H]ow often does it happen, when we fancy that we do truly know a disease, at least so far as to name it correctly, and to discriminate it from others, – that our knowledge is merely the knowledge of a name; or of phenomena not essential to the real disease; or, lastly, the belief of the existence of a condition of parts which is totally different from that which truly does exist! In this case our nosological knowledge is of very little use; or tends merely to satisfy our own minds without benefiting our patient.[21]

Latham agreed, asserting that he always had 'a great unwillingness to multiply names, especially where things are essentially the same, and differ only in being greater or less'. On this basis, auscultation could be 'greatly simplified for practical purposes'.[22] Both Forbes and Latham reorganized and reduced what they saw as categorical and linguistic excesses, reforms that were seen in nationalist terms. W. O. Markham, another English translator of auscultation texts, colourfully denounced the 'flatulent nosological vocabularies' of French doctors: 'Perhaps no one thing has more tended to impede the progress of auscultation, than the vagueness and diversity, I might say the repulsiveness, of the terms employed by different authors to indicate its phenomena'. Accordingly, Markham deemed Latham's lectures to be a shining example of 'Saxon simplicity' that divested auscultation of 'Gallic exuberances'.[23] And another review read widely in literary circles declared that Latham had put auscultation 'in an English garb of thought and language, and adapted it to English common sense'.[24] According to these readers, it was the duty of the British doctor to corral language, ensuring that it corresponded to the acoustic real, nothing more and nothing less. As Latham explained,

> I have not names for all the strange sounds which I hear every day proceeding from pulmonary cavities. And perhaps it is well that I have not. Auscultation is a new thing; and therefore we are rather disposed to make too much of trifles connected with it; a fault which it is worth your while to guard against.[25]

Yet it was precisely the power of words to multiply entities beyond necessity that authors found useful and appealing. As any author knew, giving something a local habitation and a name could bring it into being in the reader's imagination; in some sense, all of fiction relied on this principle. In fact, in one particular case involving the stethoscope, doctors and an author together (inadvertently) wrote into being a medical tool that never existed – but readers wished it had. Whether by mistake or intention, Clark's 1820 treatise introduced Laennec's invention to Britain as the 'sthenoscope' (evoking 'sthenia').[26] Perhaps 'sthenoscope' was a misremembrance or a typo, or perhaps it intentionally invoked John Brown's eighteenth-century medical theory of sthenia/asthenia (overabundance/deficiency of vital force) in yet another attempt to claim the stethoscope for Britain. Such possibilities are speculative. Whatever the case, as a result of Clark's publication (and later his prominence as Physician-in-Ordinary to Queen Victoria's household), other doctors repeated the misnomer, including Alexander Crichton, a Scottish physician to the Russian imperial family.[27] The term then took on a life of its own in fiction.

Edward Bulwer-Lytton's *Lucretia* (1846) gave 'sthenoscope' a robust afterlife, long after doctors had corrected it in medical manuals. In this first of Bulwer-Lytton's many novels depicting auscultation, one character declares that his son's 'lungs are sound enough' and 'his heart would bear the sthenoscope to the satisfaction of the College of Surgeons'.[28] The misnomer was corrected in some editions, but 'sthenoscope' persisted in numerous others on both sides of the Atlantic at least through 1896, until *Lucretia*'s audience suspected there really was a sthenoscope. In 1885, one educated reader asked *Notes and Queries* whether

> the College of Surgeons ever heard of a *sthenoscope*? [. . .] Of course some very clever people will insist that it is simply a misprint for *stethoscope*; but I do not think it is; and if there is not at present such a word, why should there not be? It is a useful word.[29]

This anecdote (albeit limited to the confines of an eclectic academic journal) suggests that readers without medical expertise sometimes relied not just upon their own encounters with doctors but upon fiction to tell them about the field, its specialized knowledge, and what doctors said or ought to say. In complement to Michael Brown's 'medico-gentility', this is also an example of what might be called 'genteel-medicine', a general interest in demonstrating medical knowledge among the readerly public.

The story of 'sthenoscope' might be aberrant, but it demonstrates how medicine and fiction could be entangled in the nineteenth century, as well as the power they had to shape each other. More importantly, it reveals a feature that was characteristic of fiction depicting auscultation. Authors availed themselves of the peculiarities of medicine – argot, professionalization, technique, prejudice – in order to heighten realism in a curious way: while British doctors sought to eliminate imaginary sounds, delimit acoustic meaning, and speak as plainly as possible about what they were hearing, authors sought to do the opposite. It was precisely auscultation's tendency to multiply entities beyond necessity that allowed authors to produce the paradoxical fictional effect of demanding that readers listen to sounds they could never hear.

Medical Professionalization in Fiction

In fiction, as in real life, stethoscopes nearly always elicited special attention from nineteenth-century readers because auscultation was often used to determine life or death. For the Victorians, the stethoscope was like a siren demanding to be listened to, with the fate of characters and plots hanging in the balance. Readers and authors were accustomed to associating auscultation with alarm in part because house calls were much more common – doctors carried stethoscopes in their top hats or overcoats to use at a moment's notice – allowing relatives, friends, and servants to witness the technique and feel the impact of its revelations in their domestic space. To take but one famous example, Elizabeth Gaskell recounted that a doctor was called to the Brontë home. Anne

> was examined with the stethoscope, and the dreadful fact was announced that her lungs were affected, and that tubercular consumption had already made considerable progress. A system of treatment was prescribed, which was afterwards ratified by the opinion of Dr. Forbes.[30]

Tuberculosis – which annihilated the Brontë siblings and was a chief killer in the nineteenth century – was a particular target of auscultation since the disease affected breathing and the lungs.[31] Stethoscopes, then, were 'maximally effective at activating [. . .] readers' existing experiential traces' not just because they were familiar but because they correlated with momentous events.[32]

For such reasons, authors dubbed the stethoscope an 'instrument of fate', or 'that little instrument which thrills us all with the aching pain of fear when we see it in the doctor's hand', as it was called by Mary Elizabeth Braddon, who included auscultation in no fewer than ten of her novels and, according to Tabitha Sparks, wrote the earlier *The Doctor's Wife* (1864) 'in an attempt to elevate her fiction to the status of realism'.[33] In fiction, auscultation almost always elicits a peak in attention, accruing significance. To put it in terms of Nicholas Dames's 'wave-theory of novelistic affect', the stethoscope triggers the 'more rigidly hermeneutic drives of suspense and revelation that create a particularly rapt, if necessarily short-lived, form of attentiveness' amid '"relaxing" subplots, or purely discursive passages'.[34] Authors deployed it this way. Auscultation, for instance, effectively draws a George Eliot novella to a close. Though much of 'Janet's Repentance' follows the machinations of two doctors in Milby, it is not until Janet resorts to soliciting an outside doctor, who 'reveal[s] by the stethoscope [. . .] the worst', that Edgar Tryan's fate is sealed and the narrative draws to a close.[35] In Eliot's first depiction of the stethoscope, auscultation ends.

In other stories, it accosts. To keep readers rapt, authors dramatized the stethoscope as a weapon and auscultation as a violent extraction of sonic truth from the body. Though partly comic, such brutal depictions heightened readerly attention and evince the sense that auscultation is a kind of Foucauldian confession. In their novels, Braddon called auscultation an 'ordeal by stethoscope'; one of George du Maurier's doctors 'punched and stethoscoped'; and William Makepeace Thackeray groused that 'a doctor will punch your chest, your liver, your heart, listen at your lungs, squeeze your pulse, and what not'.[36] James Payn compared the stethoscope to a Roman gladius and medical examination to the Inquisition in *Carlyon's Year* (1867–68):

> To see him cast himself, stethoscope in hand, upon his victim, was to witness a gladiatorial exhibition; but in reality his every movement was directed with the utmost nicety and skill. The examination ended, Dr. Throb put certain questions to the patient regarding his own symptoms, exactly as though he were himself the chief inquisitor, and Carlyon, a heretic, doomed, upon the slightest show of hesitation, to the thumbscrews, rack, and stake.[37]

The aptly named Payn later doubled down in an 1876 novel:

> the doctor applied his stethoscope, putting his head on one side, like a sagacious magpie, and listening attentively. Who has not at one time or another of his life been thumped about, and bidden to take deep breaths, to cough, and all the rest of it, with a human ear glued to his chest?[38]

As these excerpts from Payn's novels show, largely because of house calls, authors expected their readers to be familiar with the physical drama of auscultation and for it to heighten their readerly sensibilities.

Importantly, it was the *display* of auscultation – not the understanding of its principles, nor the hearing of its sounds – that readers understood, engendering a fiction all its own. Patients, whether consciously or not, sometimes *wanted* to be a bit bewildered by what doctors said and did. As laypeople, the less they understood medicalese, the more sophisticated it could seem. One character, for example, delights in a London surgeon's jargon during auscultation. The 'impassive chevalier reported on each organ in turn without moving his ear from the key-hole', muttering about 'kardiac insufficiency', 'mitral', 'tricuspid', 'hypertrophy', and 'a slight diastolic murmur'. In the ultimate act of doctorly incomprehensibility, he 'scratched a prescription which would not have misbecome the tomb of Cheops'. The mother, a 'connoisseur' of doctors, is overjoyed by his performance.[39] The very fact that these words and actions are, for her, empty signifiers reinforces her belief in the profession. As Jaffe writes, 'The desire for the real is also a desire to be told what to do'.[40]

Indeed, once the stethoscope had become the *sine qua non* of the medical profession in the nineteenth century, elevated to an emblem, it did not need to be used at all. Fiction capitalized on the public's logic that you did not have a stethoscope because you were a doctor; you were a doctor because you had a stethoscope. As today's television, movies, websites, and advertisements demonstrate, the mere appearance of a stethoscope can validate a doctor's credentials, and an attack on the tool is an attack on the profession. One of Arthur Conan Doyle's short stories, 'A False Start' (1891), exemplifies this principle. Horace Wilkinson, a recently minted physician starting his practice, knows that he must lure potential patients with a display of the exotic accoutrements they associate with the profession. Constantly rearranging his office 'so as to impress the casual patient', Wilkinson 'cleared all the little instruments out of his pocket-case – the scissors, the forceps, the bistouries, the lancets – and he laid them all out beside the stethoscope, to make as good a show as possible'.[41] Wilkinson displays the stethoscope to convince the other characters that he is a physician. Analogously, Doyle (who had medical training) displays Wilkinson displaying the stethoscope to convince readers that he is a physician *qua* character. Both Wilkinson and Doyle make the stethoscope the crucial object by which we identify the doctor's profession.

Wilkinson makes this explicit when he is summoned to treat Sir John Millbank's wife. Assuming Wilkinson is inexperienced because of his youth, Millbank grumbles that he

> won't have her thumped about all over the chest, or any hocus-pocus of the sort. She has bronchitis and asthma, and that's all. If you can cure it, well and good. But it only weakens her to have you tapping and listening; and it does no good, either.

Old Millbank, it seems, has observed plenty of auscultation at home, and – witness to display but not privy to principles – he deems it an unnecessary assault, just as the narrators in Braddon, Thackeray, and Payn do. But Wilkinson will not tolerate Millbank's amateur opinion on auscultation: 'the profession was to him a holy thing, and a flippant word about it cut him to the quick'. Millbank yields to Wilkinson's umbrage, and in 'a quarter of an hour [Wilkinson] had sounded and sifted the case to the uttermost', thereby saving Lady Millbank.[42] Wilkinson thus uses a display of auscultation to overcome Millbank's age, power, and wealth by asserting his professionalization, which negates his youth, inexperience, and lack of community standing.

As Doyle's tale indicates, deploying the specialized symbols of medicine could overcome many of the public's biases and assumptions, including those against women. Victorian medicine was so overwhelmingly and prescriptively male that a mother in one novel declares, 'There is no such thing' as a 'female physician', a 'she-doctor', a 'doctress', at least 'none in this country, happily'.[43] Nothing, of course, was further from the truth. Florence Nightingale and Mary Seacole reformed nursing; Elizabeth Blackwell became the first woman to take a medical degree in America; the Edinburgh Seven – Sophia Jex-Blake, Isabel Thorne, Edith Pechey, Matilda Chaplin, Helen Evans, Mary Anderson, and Emily Bovell – would become the first women to matriculate as undergraduates at a British university when they enrolled in 1869 to study medicine; and Elizabeth Garrett Anderson, the first woman to obtain medical qualification in Britain, helped found the London School of Medicine for Women.[44]

All of these women had knowledge of auscultation and advocated it as a central practice of the profession.[45] Denouncing the Millbanks of the world, for example, Nightingale insisted that lay bystanders were foolish to disagree with 'the opinion of the medical attendant, given, perhaps, after years of observation, after using every help to diagnosis afforded by the stethoscope, the examination of pulse, tongue, &c'.[46] And Blackwell wrote at length to a friend in a letter about how important 'accustoming [her] ear to the stethoscope' was:

> I feel that I have made progress, and detect sounds that I could not distinguish on my entrance. I advise you, E., to familiarise yourself with the healthy sounds of the chest. When you go home, auscultate all the family.[47]

Because these women received undue scrutiny and knew that they were litmus tests for women's medical aptitude, they were scrupulous in their display of professional techniques, tools, and jargon. As with Wilkinson, if you had these things, you were a doctor, regardless of gender.

Another Doyle short story, 'The Doctors of Hoyland' (1894), dramatizes this dynamic. Female physician Verrinder Smith outrages, impresses, outperforms, and jilts James Ripley, her male colleague. Though Ripley's 'whole conservative soul rose up in revolt at the idea' of a female physician, he is impressed by the '[e]laborate instruments, seen more often in hospitals than in the houses of private practitioners' in Smith's consulting room, including a 'sphygmograph [. . .] and a gasometer-like engine, which was new to Doctor Ripley'.[48] Indeed, in her inaugural address to the women's medical school, Anderson urged the real Verrinder Smiths to devote themselves to a full year of 'mastery' of all the 'machinery' of medicine, which included the 'stethoscope, ophthalmoscope, laryngoscope, sphygmograph and microscope', as well as surgical appliances, gynecological instruments, and pharmaceuticals.[49] Ripley begins to accept Smith as a member of the profession partly because he feels outdone 'by the singular, new-fashioned instruments with which she tapped, and peered, and sounded'. She ultimately proves the more dedicated physician when, 'tapping her stethoscope upon her knee', she gently rebuffs Ripley's marriage proposal in order to continue pursuing her profession, remaining unwed as Blackwell and her doctor sister Emily did.[50] Just as Wilkinson's professional knowledge of auscultation overrides cultural prejudices against his youth, so Smith's professional knowledge overrides prejudices against her gender.

Once medical authority was established – by both individuals and the professional community – doctors could wield it ruthlessly. Participating in the significant shift in diagnostic practice from patient testimony to doctor examination, from confessing subject to confessing body, from symptom to sign, Forbes insisted that in some cases, doctors 'must never trust to the reports of patients themselves or their attendants, as we are almost always sure of being misled by their prejudice or ignorance'.[51] In particular, doctors touted the stethoscope and certain auscultative techniques because they 'prevent[ed] the unarmed ear from hearing [the patient's voice] issuing from the mouth'.[52] Accordingly, a doctor in a Wilkie Collins novel 'hardly even made a pretence at listening' to his patient: 'With a frowning face he applied the stethoscope, felt the pulse, looked at the tongue – and drew his own conclusions in sullen silence', and 'he sneered as if he doubted whether [the nurse] was speaking the truth'.[53] And another Doyle short story takes medical authority to absurd heights. In it, the Prime Minister – perhaps the most powerful man in the British empire, if not the world – calls the doctor 'our medical autocrat'. He jokes with the Foreign Minister that in the company of 'a bishop one may feel at one's ease. They are not beyond the reach of argument. But a doctor with his stethoscope and thermometer is a thing apart. Your reading does not impinge upon him. He is serenely above you'.[54] Armed with a stethoscope, a doctor becomes an oracle, the ultimate wielder of truth. 'And the stethoscope never deceives?', a patient doubtfully asks in another novel. 'Never', his doctor replies.[55]

So, we have seen in this section how important the marks of medical professionalization were in nineteenth-century fiction depicting auscultation. Stethoscopes grabbed readers' attention, and auscultation was depicted as a violent extraction of truth. Such specialized tools and techniques, medical things embodying ideas, became central to recognizing doctors as such, in both real life and realistic fiction. They were powerful enough to overturn biases against youth and gender in Victorian culture and to make unusual characters acceptable to readers. Most importantly, they conditioned readers to tolerate bewildering language and activities as part of a believable medical experience. Because of this particular combination of features, auscultation invited a particular mode of reading, a particular device of realism that has its counterparts but is perhaps best highlighted in scenes depicting this medical practice. In the rest of this chapter, I formulate some of the features of this peculiar encounter with fiction.

Part 2: Subaudition, Semi-Understanding (and Mediate Reading)

According to classical tradition, Momus, the god of criticism or mockery, told Vulcan that he should have placed a lattice in the human chest so that a person's intentions could be seen plainly. In *Tristram Shandy* (1759–67), Laurence Sterne uses this myth to analogize fictional depictions of characters' interiorities:

> If the fixture of Momus's glass in the human breast, according to the proposed emendation of that arch-critick, had taken place, [. . .] nothing more would have been wanting, in order to have taken a man's character, but to have taken a chair and gone softly, as you would to a dioptrical bee-hive, and look'd in, – view'd the soul stark naked; – observed all her motions, – her machinations; [. . .] then taken your pen and ink and set down nothing but what you had seen, and could have sworn to: [. . .] But this, as I said above, is not the case of the inhabitants of this earth; – our minds shine not through the body, but are wrapt up here in a dark covering of uncrystalized flesh and blood; so that, if we would come to the specific characters of them, we must go some other way to work.[56]

This passage conflates several of our modern valences for 'character' – a person's ethics, a fictional persona, a typographical symbol – and equates the disclosure of a man's interiority with the penetration of his body, an equation perhaps intended literally in this period of phrenology, physiognomy, and humoral balancing.

In his nascent novel, Sterne's eighteenth-century dream of peering into the body and fulfilling Momus's proposal persisted well into the nineteenth century, during which medical texts gave authors reasons to think this way.[57] Forbes expressly invoked Momus in his preface to Laennec, proclaiming that auscultation had 'realized the wish of the ancient philosopher' by placing 'a window in the breast through which we can see the precise state of things within'.[58] Auscultation is a sort of aural vivisection, a way of observing the 'Motions of the Internal Parts of Bodies', as Hooke put it, without inflicting pain or death on the specimen or patient. Rothfield, discussing Balzac, Flaubert, and Zola, has likened the 'doctor:patient' to the 'novelist:character' relationship in his chapter on *Middlemarch*.[59] Menke has shown how writers such as Eliot and Lewes conceived of literature and criticism as a vivisection that lays bare the psychology of characters, real and fictional. And multiple scholars have identified *Middlemarch*'s 'other side of silence' as an invocation of auscultation's revelatory power.[60] All of this is accurate. But such scholarship has overlooked how authors valued concealment in these operations as well. Unlike vivisection, auscultation intentionally obscures as much as it reveals; its aim is to keep a surface intact even as it penetrates that surface. Its advantage consists not in wholesale disclosure but in maintaining the relationship between surface and depth, necessitating speculation and inference. After all, *Middlemarch*'s narrator asserts that we 'could hardly bear' the unfiltered 'keen' vision and feeling of human tragedy. In some sense, the fact that we are 'well wadded' – that others' 'minds shine not through the body' and 'we must go some other way' to imagine them – is what draws us into realistic fiction.[61] As Jaffe puts it, 'the inaccessibility of that capital-R "Real" keeps the idea of it in play'.[62] It is not just *display* (in this case, of authorial omniscience) on the author's behalf but also the author's *withholding* of total experience, the feeling that she hears something we cannot, that clinches a character's verisimilitude. Jaffe argues that nineteenth-century writers and subsequent critics (including those advocating 'surface reading') espousing realism depend on the visual regime, and Rothfield declares that '[e]very empirical science is devoted to the ideal of dragging the invisible into the range of the visible'.[63] But is there an alternative? Indeed, how can we explain an effect of realism that marshals the visual for the express purpose of evincing and emphasizing the acoustic?

British authors gave this paradigm of revealing-via-withholding a name – *subaudition* – or 'the action or an act of mentally supplying something that is not expressed; something that is mentally supplied or understood; an implied or understood meaning'.[64] On the one hand, subaudition may be treated as simple implication or innuendo; as with implicature or parataxis, readers who are well enough acquainted with context and culture can supply what goes unsaid in fiction. On the other hand, readers must still supply such things, must conjure them based on what they have already witnessed, things that expressly do not exist in the text. In this way, subaudition is different from mere imagination: the author conspicuously provokes the reader to fill a gap while simultaneously refusing to provide the words to do so. Authors arouse epistemology to make readers supply ontology. Subaudition therefore solicits Elizabeth Ermarth's 'consensus', but that consensus involves not just distance but blockage, an agreement to leave innards inside and let sounds go unheard, rendering language nugatory. 'Subaudition' specially resonates with auscultation because the word denotes an 'instance of barely hearing or perceiving something'; it evokes the practice of hearing through or beneath something; and its verbal and adjectival counterparts – *to subaud*, and *subaudible* – appeared (in 1826 and 1837, respectively) during auscultation's formative period when the earliest fiction depicting auscultation was being published.[65] These features make subaudition unusual and unusually linked to auscultation.

I will provide just two examples here, both from novels depicting auscultation in which authors directly appeal to subaudition, to illustrate what I mean. In *What Will He Do With It?* (1857–58), Edward Bulwer-Lytton introduces a character who speaks with a debilitating stammer, which the novelist initially represents by repeating consonants and dashes. But the narrator quickly dispenses with this typographic mimesis: 'Having thus far indicated that oral defect in our new acquaintance, the reader will cheerfully excuse me for not enforcing it overmuch. Let it be among the things *subaudita*' going forward.[66] The stammer having been illustrated, the narrator replaces it with standard speech; readers are asked to continue to 'hear' it, though it is no longer represented on the page. The implication is that the character keeps stammering 'in real life' and that the narrator has heard and hears him do it. As with auscultation, we play witness to the narrator's hearing, but we no longer hear it ourselves. The narrator's refusal to continue representing the character's stammer piques our interest in and construction of what it 'really' sounds like.

In *The Virginians* (1858), William Makepeace Thackeray invokes subaudition as a broader feature of fiction writing, similar to Laurence Sterne's pronouncements on character. As an aside to the reader, the narrator asserts that the 'real business of life [. . .] can form but little portion of the novelist's budget':

> [L]aw, stockbroking, polemical theology, linen-drapery, apothecary-business, and the like, how can writers manage fully to develop these in their stories? All authors can do, is to depict men out of their business – in their passions, loves, laughters, amusements, hatreds, and what not – and describe these as well as they can, taking the business part for granted, and leaving it as it were for subaudition.[67]

Perhaps because descriptions of characters' everyday business are too boring, too intricate, or too unknown, we are asked to subaud – to imagine, infer, and speculate, based on our real encounters with apothecaries and others – what their lives are like. In the real world, we can only partly understand others' occupations – that is, we understand enough to interact with them and use their services but not enough to provide those services ourselves. Similarly, in Thackeray's novel world, we can only partly-imagine the unexpressed lives of his characters – that is, we can imagine enough about these characters based on our real-world experiences but not enough to know what these particular characters, whom Thackeray has fabricated, do with their lives. I think it is this complex activity of understanding but not understanding enough, of imagining but not imagining enough – akin to John Plotz's attentional model of 'semi-detachment' – that Eliot describes in *Middlemarch* and that fiction depicting auscultation thrives upon.[68]

With the operations of subaudition in mind, we can re-read several passages of auscultation in fiction in which stethoscopes are misused as instead exhibiting semi-understanding that would otherwise only serve comic relief.[69] In one novel, for example, a little girl uses a doctor's stethoscope as a 'trumpet and a telescope'.[70] We deem her ludic play to be charming and naïve, but she at least knows that stethoscopes are acoustic instruments (like trumpets) that amplify sensory stimuli (like telescopes). In another novel, an Irish yokel denounces a political candidate by exclaiming, 'Be Jove, he don't know a stethoscope from an ear-trumpet!'[71] We read his outburst as a paroxysm of local colour, but he knows that the stethoscope and ear-trumpet are comparable because they operate on the same acoustic principles, even though they are obviously different objects. And in an 1871 *Punch* cartoon by George du Maurier, a doctor's daughter misuses her father's stethoscope to eavesdrop on a love interest.[72] (See Figure 1.1.)

Du Maurier's wildly popular *Trilby* (1894) depicts auscultation, and he would later help to popularize 'bedside manner' in another *Punch* cartoon, so he was no stranger to medical culture.[73] As if embroiled in a marriage plot, Angelina Squills – almost certainly an allusion to 'Dr. Squills' in Thackeray's *Vanity Fair* (1848) – steals and misuses her father's stethoscope to amplify the curate's performance (perhaps Angelina is the Trilby to the curate's Svengali). But she does so because she knows it will work; the stethoscope amplifies, whether applied to the body or a mutual wall.[74] In each of these examples, characters semi-understand: they know that the stethoscope *does* work even if they do not fully know *why* it works or *how* to use it professionally. They know this because they have witnessed displays of auscultation, though they are not privy to its expertise. In their own imaginative ways, they must therefore subaud the experience.

Figure 1.1 George du Maurier, 'Science applied to Art', *Punch* (13 May 1871), 189
Source: © Punch Limited

Subaudition, this mode of inference from semi-understanding, is heightened by two recurring tropes in fiction depicting auscultation that work in tandem: (1) auscultation's ability to render patients completely transparent to doctors; and (2) doctors' refusal to disclose what they hear while performing it. Explaining this transparency, a doctor in one Wilkie Collins novel reminds his patient that 'Science watches', as if through Momus's window: '[S]he sees you through and through. Though you are made of solid flesh and bones, and clothed, to Science you are transparent'.[75] And a doctor in one of Braddon's short stories 'sounded, rapped, and listened' to his patient's body, peering with 'a look which suggested that for him the outward semblance of a man, coat and waistcoat included, was but a glass case through which he saw the inner machinery'.[76] Authors deploy such declarations, I argue, not just to characterize doctors, but to assert that there is indeed an underlying reality accessed only by doctors. This establishes that there is a something to hear and that an authority hears it. As Dickens writes of Physician in *Little Dorrit* (1857), 'Where he was, something real was'.[77]

The second trope – doctors' refusal to disclose what they hear – produces the irony that we almost never get to hear that something. Despite the elaborate taxonomy of auscultative sounds exhaustively described in medical literature and the limitless onomatopoeia at authors' disposal, I have come across very few examples of direct description of sound, even when narration or free indirect discourse is given to a doctor. In this way, Victorian fiction presents the opposite audibility to what Melba Cuddy-Keane has called 'auscultation' in modernist narratives (especially Virginia Woolf's), in which novelists attempt to re-create actual sounds in general and the experience of hearing them.[78] In Victorian fiction, by contrast, auscultative sounds are most often not described at all or filtered through metaphors about their significance, as when one doctor hears

> a little faint sound in his polished black stethoscope, which was as terrible in its import as the noise of the loudest destroyers, as the crack of close thunder, the roar of cannon, the hiss of the hurricane, the explosion of a mine![79]

The enumerated sounds are vivid, perhaps even onomatopoeic, but they are compared to the *import* of the auscultated sound, not the sound itself; in fact, the boisterous noises are quite the opposite of the 'little faint' sound actually heard by the doctor.

Moreover, Victorian fiction emphasizes how intentionally inscrutable doctors' faces are during auscultation. A patient's mother might look on 'with greatest interest and anxiety [. . .] watching the apothecary's countenance to try to find out what opinion he was forming as he proceeded with his soundings'.[80] A father might follow 'each movement of the stethoscope and [watch] every expression of the doctor's face'.[81] But, thwarting this scrutiny, a doctor routinely 'school[s] his muscles to maintain a rigidity of expression – a calm, a repose, that nothing could disturb' when 'listening attentively'.[82] One novel even makes this opacity a point of professional pride:

> On one thing the doctor prided himself greatly, that he never let a patient know what he thought of him. He would bid a man remove his waistcoat with a semi-jocund air, and the next instant listen to a peculiar 'click' inside his frame, which betrayed the presence of heart-disease liable at any moment to carry the man off, without altering a muscle of his face or a tone of his voice.[83]

From a professional standpoint, it makes sense that a doctor would strive to remain expressionless during auscultation as a means of precluding undo anxiety or interference from laypeople (of the kind Nightingale bemoaned) before the examination is complete. This inscrutability also intensifies the scene's suspense, contributing to the peak of attention in novelistic affect that I delineated earlier. But there is another more important effect produced by drawing so much attention to a sound and then denying access to it.

To bring us full circle, back to British doctors' concerns about auscultative categories run amok, novelists rouse our faculties of subaudition because it multiplies entities beyond necessity, firing our imaginations to manufacture realities that we will believe with conviction. Novelists, in other words, encourage precisely what doctors discourage, turning the follies of auscultation into an engine of realism. A fascinating encounter in Payn's *Fallen Fortunes* (1876) illustrates this well. During a medical examination, John Dalton explains to his examiner, Dr. Curzon, that he is suffering from 'palpitations' powerful enough to cause 'an impending sense of death'. Curzon suspects that Dalton has been reading medical texts on heart-disease – a layperson meddling in professional matters – which has made him hypochondriacal. He rebukes Dalton:

> I sometimes wish that all our professional books were tabooed to the public, as the Bible used to be to laymen. They have done more harm by putting morbid fancies into people's heads than they have done good in healing their real diseases. I would not have them chained to a church table, because all the women go there, and would be sure to read them; but I would have them written in dog-Latin, so that neither the learned nor unlearned should be able to make them out, but only doctors.

A little knowledge, in other words, is a dangerous thing. Dalton retorts by insisting that what he has read 'corroborates what I have felt', to which Curzon promptly responds, 'No doubt, my dear sir; but it is astonishing how these things seem to suggest themselves, when they are, in fact, suggested'.[84] Here, Curzon outlines the process by which a semi-understanding of medicine causes laypeople to imagine diseases that are not there, and he would ban any access to professional texts to prevent it.

Just a few lines later, however, Curzon does the one thing that encourages manufacture of unrealities perhaps more than any other in auscultation novels: he pulls out his stethoscope, asserting, as we have seen doctors do before, that it is 'very truthful, and little apt even to make a mistake'. This assertion of authority sheds itself on the stethoscope with an intensity we have not yet seen, manufacturing the ultimate unreality: personification of the stethoscope. Having finished auscultating, Curzon asks,

> 'Have you had any anxieties of late, Dalton?'
>
> The doctor was still engaged in thumping when he put this inquiry, so he could not see how the other's pale face flushed.

> 'You need not reply to that question unless you please', continued the investigator, 'for I have already been informed that it is so'.
>
> 'Who can have told you, since nobody knows of them but myself?' inquired Dalton, quickly.
>
> 'The stethoscope. Now you can button up your waistcoat'.[85]

In this complex scene, the confluence of auscultation, subaudition, professionalization, and authority makes the stethoscope speak of its own volition. The stethoscope becomes not merely a metonym for professional knowledge but practically an agent. It is what Steven Connor calls a 'vocalic body', a body produced by a voice, just as voices are produced by bodies (the same principle by which doctors are produced by stethoscopes, described earlier).[86] Indeed, the personification that arose from auscultation was present long before stethoscopes appeared in novels. In his preface to Laennec, Forbes observed,

> It must be confessed that there is something even ludicrous in the picture of a grave physician formally listening through a long tube applied to the patient's thorax, as if the disease within were a living being that could communicate its condition to the sense without.

This is a personification reminiscent of Sterne's nod to homunculi in *Tristram Shandy*.[87]

But production of personhood and vocalic bodies was only one result of subaudition; auscultation was also invoked to manufacture entities that had nothing to do with internal sounds. In *The Virginians*, for example, Thackeray likens subaudition to auscultation in order to conjure a love letter that does not exist:

> Why do you not produce this letter? haply asks some sentimental reader, of the present Editor, who has said how he has the whole Warrington correspondence in his hands. Why not? Because 'tis cruel to babble the secrets of a young man's love; to overhear his incoherent vows and wild raptures, [. . .] Shall we play eavesdropper at twilight embrasures, [. . .] lay our stethoscope on delicate young breasts, and feel their heart-throbs? I protest, for one, love is sacred.[88]

But there is, of course, no letter – there never was and never will be – because Thackeray has not written it. It is only by playing 'eavesdropper', like Angelina Squills, and laying the stethoscope 'on delicate young breasts' that readers may subaud such a letter in their minds. This operation, moreover, may be extended to almost any circumstance involving revelation and concealment, freeing the stethoscope to become wholly metaphorical. Accordingly, a doctor in one Charles Kingsley novel applies a 'moral stethoscope' to determine the virtues of all new acquaintances, and a family in another novel almost a decade later wishes for a 'moral stethoscope' to discern a relative's secret intentions.[89] A mother in another novel chats with her daughter to suss out her romantic affections, 'using, so to speak, a stethoscope, to hear how Mabel's heart beat'.[90] And a Methodist preacher in yet another novel deftly extracts confessions from his pupils in most Foucauldian fashion:

> No veil of concealment could hide from their leader's piercing eye, the dark corners of their hearts where lurked their secret sins, speechless thoughts, and immature intents. His stethoscope for sounding the conscience was a most delicate and accurate instrument, which rarely failed in detecting the slightest symptom of spiritual disease.[91]

If auscultation multiplies entities, stethoscopes multiply just as quickly to detect them.

We have seen that the manner in which auscultation was imported from France made it peculiarly British, but we have also seen that depictions of auscultation in British novels were not merely historical tableaux, and the stethoscope was not simply a prop. Instead, they manifested a peculiar operation of realism, adding to the techniques that novelists could use to evince the real. Indeed, while we think of the stethoscope now primarily in terms of listening, the nineteenth century also thought of it in terms of writing. Laennec's treatise, after all, explained how the very first stethoscope was fashioned out of a 'quire of paper', how 'pencil cedar' was the best construction material, and how the stethoscope must remain sealed upon the body during auscultation, 'which can only be done by attending to the author's direction of holding it as we do a pen'.[92] Perhaps, then, fiction depicting auscultation not only calls for us to re-read but for us to rewrite how we read it altogether. In addition to close, surface, distant, or even symptomatic reading, in other words, perhaps we need a formulation of *mediate* reading, in the spirit of auscultation.

Notes

1 René-Théophile-Hyacinthe Laennec, *A Treatise on the Diseases of the Chest*, trans. John Forbes (London: T. and G. Underwood, 1821), pp. 284–5.

2 See Jonathan Sterne, *The Audible Past: Cultural Origins of Sound Reproduction* (Durham, NC: Duke University Press, 2003), pp. 90–8.

3 Michael Brown, *Performing Medicine: Medical Culture and Identity in Provincial England, c.1760–1850* (Manchester: Manchester University Press, 2018), p. 9. During the nineteenth century, there were numerous terms for those participating in medical practices: 'physician', 'surgeon', 'apothecary', 'medical man', 'general practitioner', 'family doctor', and many more. Brown

uses the blanket term 'medical practitioner'. In this chapter, however, unless I have reasons to call them otherwise, I opt for Christopher Lawrence's policy: 'For stylistic reasons I will refer to regular practitioners of all sorts as doctors, even though for much of the period under consideration they were not accorded that courtesy title' (*Medicine in the Making of Modern Britain*, 1700–1920 (London: Routledge, 1994), p. 5).

4 See Roland Barthes, *The Rustle of Language*, trans. Richard Howard (Berkeley: University of California Press, 1986), pp. 141–8; Stephen Best and Sharon Marcus, 'Surface Reading: An Introduction', *Representations*, 108.1 (2009), 1–21; Leah Price and Matthew Rubery (ed.), *Further Reading* (Oxford: Oxford University Press, 2020).

5 Lawrence Rothfield, *Vital Signs: Medical Realism in Nineteenth-Century Fiction* (Princeton: Princeton University Press, 1992); Richard Menke, 'Fiction as Vivisection: G. H. Lewes and George Eliot', *ELH*, 67.2 (2000), 617–53; John Picker, *Victorian Soundscapes* (Oxford: Oxford University Press, 2003); Michel Foucault, *The Birth of the Clinic: An Archaeology of Medical Perception*, trans. A. M. Sheridan (London: Tavistock, 1973), pp. 149–73.

6 Leah Price, *The Anthology and the Rise of the Novel: From Richardson to George Eliot* (Cambridge: Cambridge University Press, 2000), p. 5.

7 Ian Watt, *The Rise of the Novel: Studies in Defoe, Richardson and Fielding* (London: Chatto & Windus, 1957); Barthes, *Rustle of Language*; Elizabeth Deeds Ermarth, *Realism and Consensus in the English Novel: Time, Space and Narrative* (Edinburgh: Edinburgh University Press, 1998); Elaine Auyoung, *When Fiction Feels Real: Representation and the Reading Mind* (New York: Oxford University Press, 2018).

8 Ann Elizabeth Gaylin, *Eavesdropping in the Novel from Austen to Proust* (Cambridge: Cambridge University Press, 2002), p. 24.

9 Elaine Freedgood, *Worlds Enough: The Invention of Realism in the Victorian Novel* (Princeton: Princeton University Press, 2019), p. xii.

10 *OED*: 'realistic, adj.'.

11 Audrey Jaffe, *The Victorian Novel Dreams of the Real: Conventions and Ideology* (Oxford: Oxford University Press, 2016), p. 8.

12 Ibid., p. 148.

13 James Africanus Beale Horton, *The Diseases of Tropical Climates and Their Treatment* (London: J. and A. Churchill, 1874). The other seven Nigerians were William Davies, Obadiah Johnson, Nathaniel King, Akinsiku Leigh-Sodipe, Orisadipe Obasa, John Randle, and Oguntola Sapara. During this time, Kadambini Ganguly and Anandi Gopal Joshi also became the first Indian women to obtain British medical degrees.

14 Samuel Warren, *Ten Thousand A-Year*, 3 vols (London: William Blackwood and Sons, 1841), III, p. 264.

15 George Eliot, *Middlemarch* [*Cabinet Edition*], 3 vols (London: William Blackwood and Sons, 1878 [1871–72]), II, pp. 21, 228.

16 Robert Hooke, *The Posthumous Works* (London: Sam. Smith and Benj. Walford, 1705), p. 39.

17 Picker, *Victorian Soundscapes*, pp. 6, 82–109, 152–53n13; W. F. Bynum, *Science and the Practice of Medicine in the Nineteenth Century* (Cambridge: Cambridge University Press, 1994), pp. 40–41; Eliot, *Middlemarch*, I, pp. 297–8.

18 Charles James Wills, *The Pit Town Coronet: A Family Mystery*, 3 vols (London: Ward and Downey, 1888), II, p. 64. Wills was a doctor. Arthur Conan Doyle, 'Behind the Times', in *Round the Red Lamp* (London: Methuen, 1894), p. 4.
19 Rudyard Kipling, *Rewards and Fairies* (London: Macmillan, 1910), pp. 92, 104.
20 Adam Raciborski, *An Elementary Treatise on Auscultation and Percussion, Or the Application of Acoustics to the Diagnosis of Diseases, With a Synoptic Table*, trans. Minturn Post (New York: Collins, Keese & Co., 1839). Post graduated from the medical department of the University of Pennsylvania in 1832.
21 Forbes, 'Translator's Preface', in Laennec, *Treatise* (1821), p. xv.
22 Peter Mere Latham, *Lectures on Subjects Connected with Clinical Medicine* (London: Longman, Rees, Orme, Brown, Green, and Longman, 1836), pp. 200, v.
23 Markham, 'Translator's Preface', in Joseph Škoda, *A Treatise on Auscultation and Percussion*, trans. William Orlando Markham (London: Highley and Son, 1853), pp. ix–x, xiv.
24 [Anon.], 'Art. X. – Lectures on Subjects connected with Clinical Medicine. By P.M. Latham', *Quarterly Review*, 57 (1836), 199–223 (p. 223).
25 Latham, *Clinical Medicine*, p. 305.
26 Clark writes, 'Dr. Laennec was indisposed, but his colleague Dr. Cayol pointed out to me several cases where the information gained by the application of the *sthenoscope* was useful in ascertaining the nature of the disease' (James Clark, *Medical Notes on Climate, Diseases, Hospitals, and Medical Schools, in France, Italy, and Switzerland* (London: T. and G. Underwood, 1820), p. 153). He mentions the 'sthenoscope' twice more on p. 158.
27 Crichton uses 'sthenoscope' twice, though the errata correct both instances to 'stethoscope' (Alexander Crichton, *Practical Observations on the Treatment and Cure of Several Varieties of Pulmonary Consumption* (London: Lloyd and Son, 1823), pp. 87, 185, [262]).
28 Edward Bulwer-Lytton, *Lucretia*, 3 vols (Leipzig: Bernhard Tauchnitz, 1846), I, p. 284.
29 J. B. Fleming, 'Sthenoscope', *Notes and Queries*, s6-XI (1885), 428–9 (pp. 428–9).
30 Elizabeth Cleghorn Gaskell, *The Life of Charlotte Brontë*, 2 vols (London: Smith, Elder & Co., 1857), II, p. 93. 'Dr. Forbes' is John Forbes, discussed throughout this chapter.
31 See Anne Hardy, *The Epidemic Streets: Infectious Diseases and the Rise of Preventive Medicine, 1856–1900* (Oxford: Clarendon Press, 1993).
32 Auyoung, *When Fiction Feels Real*, p. 13.
33 William Gorman Wills, *Notice to Quit*, 3 vols (London: Hurst and Blackett, 1861), III, p. 315; Mary Elizabeth Braddon, *All Along the River*, 3 vols (London: Simpkin, Marshall, Hamilton, Kent & Co., 1893), II, p. 58; Tabitha Sparks, *The Doctor in the Victorian Novel: Family Practices* (London: Routledge, 2009), p. 63. Braddon claimed *The Doctor's Wife* was an anglicization of Gustav Flaubert's *Madame Bovary* (1856), generally regarded as an archetype of French realism. On realism in Anglo-French medical culture, see Rothfield, *Vital Signs*.

34 Nicholas Dames, 'Wave-Theories and Affective Physiologies: The Cognitive Strain in Victorian Novel Theories', *Victorian Studies*, 46.2 (2004), 206–16 (p. 214).
35 George Eliot, *Scenes of Clerical Life*, 2 vols (London: William Blackwood and Sons, 1858), II, p. 373.
36 Braddon, *All Along the River*, II, p. 58; George du Maurier, *Trilby* (London: Osgood, McIlvaine & Co., 1895), p. 206; William Makepeace Thackeray, *The Adventures of Philip*, 3 vols (Leipzig: Bernhard Tauchnitz, 1862), I, p. 205.
37 James Payn, *Carlyon's Year*, 2 vols (London: Bradbury, Evans, & Co., 1868), II, pp. 253–4.
38 James Payn, *Fallen Fortunes*, 3 vols (London: Tinsley Brothers, 1876), I, p. 243.
39 Charles Reade, *Hard Cash*, 3 vols (Leipzig: Bernhard Tauchnitz, 1864), I, p. 66.
40 Jaffe, *Victorian Novel Dreams of the Real*, p. 149.
41 Doyle, *Round the Red Lamp*, p. 70.
42 Ibid., pp. 84–6.
43 Reade, *Hard Cash*, p. 67. The mother, Mrs. Dodd, is responding to her daughter Julia, who has requested a female doctor perhaps partly because of the social complexities involved in a male doctor's inspecting her body. As Jonathan Sterne points out, mediate auscultation aided nineteenth-century propriety because a male doctor no longer had to place his ear against a woman's breast(s) (Sterne, *Audible Past*, pp. 104–17).
44 This list of 'firsts' must acknowledge Dr. James Miranda Barry, a trans man who matriculated at Edinburgh in 1809, whose gender identity has been a subject of debate in recent years. The (rather outmoded) entry on Barry in the *Oxford Dictionary of National Biography* posits that 'Barry became not only the youngest but technically the first woman in Great Britain to graduate in medicine, an honour usually accorded to Elizabeth Garrett Anderson, who graduated in 1865, the year of Barry's death' (Sydney Brandon, 'Barry, James (c. 1799–1865), army medical officer and transvestite' in *Oxford Dictionary of National Biography*: https://doi-org.laneproxy.stanford.edu/10.1093/ref:odnb/1563 [accessed 31 December 2020]).
45 Pechey provides one glimpse into the emphasis on auscultation in medical education: 'To the kindness of this eminent physician [G. W. Balfour], and to his thorough painstaking teachings, I am indebted for all that I know of diseases of the heart, and all my subsequent experience has served to bring back vividly to my mind the minutely accurate descriptions of the symptoms and physical signs revealed by auscultation which his students were privileged to hear in the Edinburgh Infirmary' (Mary Edith Pechey, *Upon the Constitutional Causes of Uterine Catarrh [thesis]* (Brighton: M. Sickelmore, 1877), p. 13).
46 Florence Nightingale, *Notes on Nursing* (London: Harrison, 1859), p. 55.
47 Elizabeth Blackwell, *Pioneer Work in Opening the Medical Profession to Women: Autobiographical Sketches* (London: Longmans, Green, and Co., 1895), p. 172.
48 Doyle, *Round the Red Lamp*, pp. 300–1.
49 Elizabeth Garrett Anderson, *London School of Medicine for Women, Inaugural Address* (London: H. K. Lewis, 1877), p. 8.
50 Doyle, *Round the Red Lamp*, pp. 304, 314.

51 Forbes, 'Translator's Preface', in Laennec, *Treatise* (1821), p. xvii.
52 René-Théophile-Hyacinthe Laennec, *A Treatise on the Diseases of the Chest*, 3rd edition, trans. John Forbes (New York: Samuel S. and William Wood, 1838), p. 42.
53 Wilkie Collins, *Blind Love*, 3 vols (London: Chatto & Windus, 1890), III, pp. 58–9. Finished by Walter Besant.
54 Doyle, *Round the Red Lamp*, pp. 182–3.
55 Payn, *Carlyon's Year*, I, p. 305.
56 Laurence Sterne, *The Life and Opinions of Tristram Shandy, Gentleman*, 9 vols (York: Ann Ward, 1760), I, pp. 165–8.
57 Michael Brown has also noted Sterne's influence on the writing style of doctors, such as the York surgeon-apothecaries, Charles and James Atkinson (Brown, *Performing Medicine*, pp. 70–5).
58 Forbes, 'Translator's Preface', in Laennec, *Treatise* (1821), p. xiv.
59 Rothfield, *Vital Signs*, p. 86.
60 See, for example, Menke, 'Fiction as Vivisection', pp. 635–66; and Picker, *Victorian Soundscapes*, pp. 7–11.
61 Eliot, *Middlemarch [Cabinet Edition]*, pp. 297–8.
62 Jaffe, *Victorian Novel Dreams of the Real*, p. 19.
63 Ibid., pp. 1–19; Rothfield, *Vital Signs*, p. 96.
64 *OED*: 'subaudition, n.'.
65 *OED*: 'subaudition, n.'; 'subaud, v.'; 'subaudible, adj.'.
66 Edward Bulwer-Lytton, *What Will He Do With It?*, 4 vols (Leipzig: Bernhard Tauchnitz, 1857), I, p. 282.
67 William Makepeace Thackeray, *The Virginians*, 3 vols (Leipzig: Bernhard Tauchnitz, 1859), III, pp. 129–30.
68 John Plotz, *Semi-Detached: The Aesthetics of Virtual Experience since Dickens* (Princeton: Princeton University Press, 2017), p. 1.
69 I proceed with 'semi-understanding', instead of 'part-understanding', because it resonates with 'semiotic', relating to both the medical study of symptoms and the linguistic study of signs, and with Plotz's 'semi-detachment'.
70 Charlotte Mary Yonge, *The Daisy Chain*, 2 vols (London: John W. Parker and Son, 1856), I, p. 251.
71 Arthur Hallam Elton, *Herbert Chauncey*, 3 vols (London: Smith, Elder and Co., 1860), III, p. 241.
72 George du Maurier, 'Science Applied to Art', *Punch*, 60 (13 May 1871), 189.
73 'Bedside manner' dates to at least the 1860s, but du Maurier's 1884 cartoon, 'Annals of a Winter Health Resort', was one of its most famous uses. *Punch*, 86 (15 March 1884), p. 121.
74 *Punch* reproduced this cartoon for a 1976 recaptioning contest. The winning entry – '"For your penance, one Hail Mary and four verses of The Lord's My Shepherd"' – was a quip that, during Foucault's lifetime, emphasized auscultation's confessional nature (P. J. Wiehl, 'Caption Competition – 376', *Punch* (13 October 1976), 658).
75 Collins, *Blind Love*, p. 74.
76 Braddon, 'His Oldest Friends', in *All Along the River*, III, p. 228. Conflations of meanings of 'case' – as in 'glass case' or 'case study' – are richly complex

in auscultation fiction. In Doyle's 'The Case of Lady Sannox', for example, a surgeon examines a woman he believes to be Turkish and whose body is almost entirely covered with a 'yashmak and veil'. 'But the surgeon was not thinking about the yashmak', Doyle writes. 'This was no longer a woman to him. It was a case' (*Round the Red Lamp*, p. 169).

77 Charles Dickens, *Little Dorrit* (London: Bradbury and Evans, 1857), p. 531.

78 See Melba Cuddy-Keane, 'Virginia Woolf, Sound Technologies, and the New Aurality', in *Virginia Woolf in the Age of Mechanical Reproduction*, ed. Pamela L. Caughie (London: Garland, 2000), pp. 69–96; Cuddy-Keane, 'Modernist Soundscapes and the Intelligent Ear: An Approach to Narrative through Auditory Perception', in *A Companion to Narrative Theory*, ed. James Phelan and Peter J. Rabinowitz (Oxford: Blackwell, 2005), pp. 382–98.

79 Philip Gilbert Hamerton, *Wenderholme*, 3 vols (London: William Blackwood and Sons, 1869), III, p. 43.

80 Marmion Wilme Savage, *Reuben Medlicott*, 3 vols (London: Chapman and Hall, 1852), I, p. 229.

81 Anna Maria Hopton, *Dawn and Twilight*, 2 vols (London, John Henry and James Parker, 1858), II, p. 328.

82 Edward Dutton Cook, *Leo*, 3 vols (London: Smith, Elder and Co., 1863), I, pp. 256–7.

83 Edmund Yates, *Broken to Harness*, 3 vols (London: John Maxwell and Company, 1864), II, p. 108. 'Click' in this excerpt is the only obvious instance of onomatopoeia I have found in nineteenth-century British fiction depicting auscultation.

84 Payn, *Fallen Fortunes*, I, p. 242–3.

85 Ibid., p. 243.

86 Steven Connor, *Dumbstruck: A Cultural History of Ventriloquism* (Oxford: Oxford University Press, 2000), pp. 35–43.

87 Forbes, 'Translator's Preface', in Laennec, *Treatise* (1821), p. xiv. Using language similar to Forbes's, Laurence Sterne writes, 'The Homunculus, Sir, in how-ever low and ludicrous a light he may appear, in this age of levity, to the eye of folly or prejudice; – to the eye of reason in scientific research, he stands confess'd – a BEING guarded and circumscribed by rights' (Sterne, *Tristram Shandy*, I, p. 4).

88 Thackeray, *Virginians*, III, p. 241.

89 Charles Kingsley, *Two Years Ago*, 3 vols (Cambridge: Macmillan and Co., 1857), I, p. 139; Lewis Hough, *William Bathurst*, 3 vols (London: Hurst and Blackett, 1865), II, p. 37.

90 Richard Ashe King, *Love the Debt*, 3 vols (London: Smith, Elder, & Co., 1882), III, pp. 194–5.

91 Luke Wesley Church, *From the Plough to the Pulpit*, 3 vols (London: Tinsley Brothers, 1874), I, pp. 167–8.

92 Laennec, *Treatise* (1838), p. 784; Forbes, 'Translator's Preface', in Laennec, *Treatise* (1821), p. xxi.

2 'Redemption From Probable Destruction'

Deafness, Isolation, and Identity in the *Autobiography* of Harriet Martineau

Clare Walker Gore

To readers familiar with Harriet Martineau's highly successful and consistently controversial literary career, her 'Letter to the Deaf' (1834) is likely to come as something of a surprise. Published after she had already burst onto the London literary scene with her best-selling series, *Illustrations of Political Economy*, the essay claims to draw on her personal experience, promising to share the strategies which have transformed her own deafness 'from being an almost intolerable grievance, [into] so much less of one to myself and my friends, than such a deprivation usually is'.[1] Yet the strategies Martineau goes on to recommend are almost entirely based on self-effacement and self-denial, as she urges the 'fraternity' of the deaf to assimilate, accommodate, and subordinate themselves to hearing society in every possible way. The ideal deaf subject she constructs is selfless to the point of masochism, cheered by the 'thrill of delight which arises during the ready agreement to profit by pain' and so absolutely committed to the principle of giving 'the least possible pain to others' that their own feelings must be continually repressed, as they observe the 'little rule of turning every sigh into a smile'.[2] While insisting that deaf people must continue to go into 'Society', Martineau suggests that 'it must be sought [. . .] under a bondage of self-denial', warning that '[w]e must submit to be usually insignificant, and sometimes ridiculous.'[3] The promised compensation is the admiration and approval that will be won from hearing friends and relatives, and the moral elevation which will provide 'cheerfulness beyond the reach of circumstance.'[4]

The contrast between this imagined deaf subject and Martineau herself could hardly be starker. By 1834, Martineau was already famous (or notorious) for her eagerness to tackle inflammatory – and conventionally 'masculine' – subjects in print, from the 'population question' to the New Poor Law, and she went on to espouse controversial causes as various as Abolitionism, women's rights, mesmerism, and atheism. Looking at her career as a whole, it is difficult not to conclude that she courted controversy rather than aiming to please, and this is in fact strongly borne out by

DOI: 10.4324/9780429291616-3

her self-representation in her later and more extended piece of life-writing, her two-volume *Autobiography* written in 1855 and published posthumously in 1877. There, she is surrounded not by 'tact and kindness', as in the 'Letter', but by uncomprehending and insensitive, if well-meaning, relatives: 'I believe my family would have made almost any sacrifice to save me from my misfortune; but not the less did they aggravate it terribly by their way of treating it'.[5] So strong is her sense of childhood suffering and alienation, in this account, that she passionately identifies with the rebellious and mistreated heroine of Charlotte Brontë's *Jane Eyre*: 'When I read it, I was convinced that it was by some friend of my own, who had portions of my childish experience in his or her mind'.[6] Like the fictional Jane, Martineau left her unhappy home to make an independent life for herself in London and beyond, and while she did live with her mother and aunt for some periods of time, she makes it very clear in her *Autobiography* that she was much happier alone. Far from submitting to being taught by suffering, and constantly moulding herself to the needs of those around her, the Martineau of the *Autobiography* is counter-culturally self-assertive and self-reliant, explicitly rejecting others' misjudgements of her capabilities in order to forge an independent literary career and an independent life.

It is in this respect, however, that a continuity between the self-writing of the 'Letter' and of the *Autobiography* emerges, and it is rooted in Martineau's construction of deafness. In both texts, Martineau insists that the deaf cannot accept advice from hearing people but must consider themselves as set apart, rendered exceptional by their deafness. In the 'Letter', this is primarily cast as a form of unwelcome isolation, and a source of additional suffering, but it is also used to justify a kind of self-reliance and self-assertion which otherwise sits awkwardly with the calls to endless patience and forbearance: 'Advice must go for nothing with us in a case where nobody is qualified to advise'.[7] As Martha Stoddard Holmes astutely points out, while the 'Letter' constructs deafness 'as an experience of loss', it also creates a deaf person who is in total control of that experience, 'a person who asks for nothing' through adopting 'a plan of rigid self-sufficiency'.[8] I would argue that Martineau actually goes beyond 'asking for nothing', in fact asking that others respect her (and other deaf people's) absolute right to judge for themselves and to be considered exceptions to ordinary rules. The anecdotes she recounts in which she made sacrifices for others' sake all seem to involve her contradicting other (hearing) people and overruling their wishes, ignoring 'reproaches from kind friends' and acting 'in opposition to much entreaty'.[9] For all that she is urged to accommodate herself to others' needs, the deaf subject of the 'Letter' ends up being cast not only as exceptional, made 'either very weak or very strong' by 'uncommon circumstances', but as heroically beyond the pale of others' understanding or – crucially – management.[10] In this chapter, I will argue that this is exactly how Martineau uses her deafness in her *Autobiography*: to make of herself a heroine who cannot be judged by others but who must be allowed to go her own way as an exceptional and necessarily independent figure.

The idea of Martineau as exceptional runs through much critical writing about her – unsurprisingly, since she strayed so far from the conventionally 'feminine' in her choice of subjects (political economy, metaphysics, the lives of revolutionary leaders),[11] and was so often treated in the press as aberrant on that very account.[12] Dorothy Mermin, for example, suggests that as 'the model, for good or ill, of an androgynous intellectual', Martineau 'was more likely to be admired than imitated' by the women writers who followed her.[13] Linda Peterson argues that this extends to the form of her life-writing, which is 'exceptional rather than representative', since by giving her *Autobiography* the form of a spiritual autobiography, Martineau essentially took a male literary form and made it her own.[14] The idea of Martineau as an exception to every rule has been challenged in recent years, with Peterson subsequently arguing that Martineau's early writings had much in common with the educational tracts published by earlier women writers, and with Susan F. Bohrer suggesting that Martineau sought to cast herself 'as an example, rather than extraordinary'.[15] Yet the image of Martineau as a woman set apart has proved hard to dislodge, and I argue that this is because it is central to her compelling *Autobiography*, in which Martineau constructs herself as both necessarily and fortuitously isolated by her deafness. Where the 'Letter' insists on the need for deaf people to 'tak[e] our place in society', the *Autobiography* repeatedly stages scenes of withdrawal and isolation as conducive to work, and of alienation and estrangement from the good opinions of family, friends, and readers as essential to growth as a writer.[16]

Martineau's understanding of deafness as necessarily isolating may strike a contemporary reader as idiosyncratic, even perverse: after all, she addresses herself to an imagined community of deaf readers in her 'Letter to the Deaf', and she was writing at a time when signed languages had long existed and were widely used by Deaf communities in Britain. However, it is rooted in two powerfully prevalent cultural attitudes of the early Victorian period. Firstly, Martineau understands deafness as 'affliction', calling it in the 'Letter' 'a great disability' and 'a great privation',[17] and a dominant way of understanding disability of all kinds in this period was as a form of isolation from 'normal' society.[18] This cultural construction of disability dominates nineteenth-century fiction, with characters as various as Eliot's Philip Wakem ('always solitary'), Dickens's Jenny Wren ('I couldn't play'), or Collins's Rosanna Spearman ('all alone') all defined as set apart from their communities; in her own novel, *Deerbrook* (1839), Martineau depicts the 'solitary' crippled governess Maria Young as defined by her isolation.[19] It is true, of course, that some fictional plotlines make space for disabled characters at their heart, and Talia Schaffer has convincingly argued that in some cases disabled characters create 'networks of care' rather than isolation.[20] But Martineau seems to have been sceptical about the possibility of disability fostering rather than fraying relations between the individual and their community. In her *Life in the Sick Room* (1844), she treats invalidism as a fundamentally lonely condition, in which the sufferer may be elevated by her experience but is also inevitably set apart by it: 'If formerly we associated with equals, it cannot be but that we must now see further than they'.[21]

Secondly, Martineau's construction of deafness as isolating can be traced to the climate of mistrust which increasingly surrounded signing deaf communities, amidst eugenically inflected fears of 'deaf clannishness' and the growing power of the oralist movement, which demanded that deaf people integrate into hearing society on terms necessarily disadvantageous to them. As Lennard J. Davis explains,

> [w]hile the audist establishment initially constructed the deaf person as a model inhabitant of the Enlightenment, a citizen in the world of print culture, it came to see deaf people, particularly those using the 'foreign' sign language, as an ethnic minority with its own history and language that must be incorporated into the state and nation.[22]

The 'otherness' of the deaf and their loneliness without strenuous attempts at integration were therefore increasingly stressed throughout the nineteenth century.

Martineau's 'Letter' and *Autobiography* were both written before the high point of the oralists' success in driving sign language out of classrooms, and she does not explicitly refer to sign language in either piece, but the influence of oralists' ideas is evident. Jennifer Esmail points out that Martineau 'was generally dismissive of the abilities of signers', citing her disparaging remarks about sign language in *Household Education* (1848), which are certainly consistent with her insistence in the 'Letter' that the partially deaf (like herself) must make ceaseless efforts to preserve and enhance their hearing through prosthetic devices and accommodate themselves to their hearing companions.[23] Indeed, the almost phobic vehemence with which Martineau insists that she has nothing in common with those who are deaf from birth and disparagement even of the partially deaf as 'often unscrupulous about truth, inquisitive, irritable, or morose' suggest that Martineau internalised many of the most negative stereotypes about deafness prevalent in her period.[24] Although she does anticipate the now widely accepted idea that disability is fundamentally relational (the 'social model' of disability), the way she deals with this is not by advocating for the re-configuring of social relations or seeking out social connection with other deaf people but by casting sociability as threatening and dangerous in itself and isolation as a preferable state, conducive to creativity and intellectual endeavour. In other words, her alignment of deafness and isolation arises from her acceptance of some of the most limiting prevalent stereotypes concerning the deaf – but she manages to configure that isolation in a way that enables her self-construction as a heroic woman of letters.

'I Gathered Myself Up for a Gallant Breasting of My Own Destiny': Deafness and the Emergent Self in Childhood and Youth

Martineau's depiction of her childhood is one of the most vivid parts of the *Autobiography*, and also by far the bleakest. Although she grew up in materially comfortable and relatively stable circumstances, as part of a numerous, middle-class family in Norwich, Martineau portrays her childhood as almost unrelievedly miserable; at one point, she laconically notes that '[b]eing usually very unhappy, I was constantly longing for heaven, and seriously, and very frequently planning suicide in order to get there'.[25] In this account, Martineau's unhappiness arises not so much from any deliberate unkindness on the part of her relatives as from their total lack of understanding. Her 'unbounded need of approbation and affection' means that their inability to sympathise with her, and frequent disapproval of her phobias and anxieties, cause her acute mental suffering.[26] These 'panics' arise from a mixture of frightening fantasies – 'I could never cross the yard to the garden without [. . .] fearing to look behind, because a wild beast was after me' – and of sensory experiences which she cannot properly process.[27] Unable to smell or taste, Martineau is tormented by the senses she does possess; phenomena as apparently harmless as light coming in through the windows, which 'seemed to advance till it pressed upon my eyeballs', or the sound of feather beds being cleaned by beating, 'a dull shock [which] used to make my heart stand still', terrify the child Harriet, who depicts herself as agonisingly vulnerable.[28]

Martineau cites ill health as the cause of this excessive sensitivity, explaining that her family believed that being starved as a baby by inadequate wet-nursing had caused her 'bad health during my whole childhood and youth, and even my deafness'.[29] Crucially, however, she also casts ill health and her subsequent loss of hearing as consequences rather than causes of her hyper-sensitivity not only to sense impressions but to others' feelings. In a peculiarly charged anecdote about her first real friendship, she explains to the reader that her friend 'E.' was 'very lame', and had to have her leg amputated, and that this had a drastic effect on her own health:

> It turned my imagination far too much on bodily suffering [. . .] I am sure that my nervous system was seriously injured, and especially that my subsequent deafness was partly occasioned by the exciting and vain-glorious dreams that I indulged in for many years after my friend E. lost her leg.[30]

Martineau's matter-of-fact tone encourages the reader to accept as obvious the startling idea that disability might be a direct consequence of imaginative identification with another's – completely different – experience of impairment. In this account, the damage done by her sympathetic identification with E.'s pain is irreparable, her deafness 'seriously aggravated', before being 'severely increased' by an unspecified 'sort of accident'.[31]

Dorothy Mermin convincingly argues that this episode is used by Martineau to illustrate the dangers of imagination and that it explains why 'Martineau minimized the role of imagination in her adult [literary] work'.[32] However, the dangers of sociability are equally invoked: this is the first time we have seen Martineau associate with a friend outside her own family (from whom she generally depicts herself as alienated), and the association leads to disaster. Having already traced much of her misery to her excessive dependence on the affection and approval of others, Martineau here asks us to imagine the self as dangerously porous to outside influences, the body itself re-made by too close an identification with another's experience. Just as she will later suffer physical collapse and debilitating ill health when she forces herself to live with her mother and is worn down by her emotional demands, here, enforced closeness to another child wrecks her health. Moreover, although Martineau insists that it was a 'wholesome discipline' to experience the 'long self-denial' of standing apart with E. rather than playing with the other children, resentment at the curtailment of her own freedoms is allowed to surface through passive-aggressive commentary. In parentheses she notes that she worried about upsetting E., because *her* feelings were 'always so tenderly considered'; the contrast with her own treatment embitters the passage.[33]

However, this episode also introduces the possibilities opened up by disability as well as the dangers of intimacy with others. Whilst deprecating her morbid fantasies, Martineau also suggests that she was made more stoical, more unusual, by her identification with E., which 'occasioned or strengthened a power of patience under pain and privation which was not to be looked for in a child so sensitive and irritable by nature'.[34] This is the first suggestion that the onset of her own disability – deafness – will make Martineau the heroine of her own life story, transforming her from the vulnerable, dependent, and even cowardly child that she shows herself to have been into the enterprising, independent heroine she becomes. When she reassures the reader that she and E. have re-established their friendship in adulthood, she notes that after she had published her 'Letter to the Deaf', 'we seemed to be brought nearer together by our companionship in infirmity'.[35] Now that Martineau, too, is disabled, they can finally meet as equals; where once Martineau resented having to stand aside with E., now they are both set apart. Tellingly, E. is cited as 'one of the most favourable influences I had the benefit of after taking myself in hand for self-government'.[36] The onset of disability and the beginning of 'self-government' are thus brought together, so that the result of excessive identification with others – deafness – actually leads to a pleasurable independence from them.

One of the frightening things about the onset of deafness in Martineau's account is that it seems definitively to preclude her ever winning the sympathy and approval from others which have so far largely eluded her. Martineau has already depicted herself as desperate for attention, the least favoured member of the family; now, her brother tells her the cautionary tale of a deaf lady he knows, 'a sort of provincial blue-stocking in her time', who makes herself 'irksome and absurd' by constantly asking that things be repeated to her.[37] The threat that this is Martineau's future is implicit. It informs the 'resolution which I made and never broke, – never to ask what was said', and it is made still more painful by the fact that the only other deaf woman she knows, 'a distant relation', is also judged to be 'far from agreeable' and avoided by the Martineau children, 'less for that, than on account of the trouble of her deafness'.[38] This time, the connection to Martineau herself is explicit:

> When I was growing deaf, all this came back upon me; and one of my self-questionings was – "Shall I put people to flight as —— —— does? Shall *I* be dreaded and disliked in that way all my life?" The lot did indeed seem too hard to be borne.[39]

The deprivation of hearing loss in itself hardly features in the *Autobiography*; as mentioned earlier, sensory experience features here almost entirely in terms of trauma, and there is no equivalent to the mournful listing in the 'Letter' of the order in which different kinds of birdsong gradually slip away.[40] Instead, the pain of deafness is in the isolation from others it engenders, the apparent impossibility of being both deaf and socially acceptable.

Yet in Martineau's account, dependence on others' good opinion has only brought unhappiness, and she is already the odd one out in her family and community on account of her awkwardness, her bookishness, and her methodical turn of mind. In effect, her deafness simply pushes her existing isolation, as it pushes her existing sensory deprivation and weak health, to a crisis point which actually marks the end of the misery which has characterised her life up to this point. Her family's failure to understand her, in particular, is taken to an extreme which is cast as redemptive: truly isolated by their readiness 'to inflict frightful pain [. . .] to relieve their own weakness and uneasiness', Martineau is finally liberated from dependence on their opinion and need for their sympathy:

> This was hard discipline; but it was most useful in the end. It showed me that I must take my case into my own hands; and with me, dependent as I was upon the opinion of others, this was redemption from probable destruction. Instead of drifting helplessly as hitherto, I gathered myself up for a gallant breasting of my own destiny; and in time I reached the rocks where I could take a firm stand. I felt that here was an enterprise; and the spirit of enterprise was roused in me; animating me to sure success, with many sinkings and much lapse by the way.[41]

Martineau's dramatic turn of phrase makes it clear that this is a turning point in her narrative, invoking both the salvation scene of a spiritual memoir (with its biblically inflected metaphor of 'the rocks where I could take a firm stand' and new promise of 'redemption'), and the language of the liberal individualism and free-market economics which were to become the cornerstone of her political commitments and of her first literary success: 'the spirit of enterprise' makes Martineau sound like a capitalist hero rather than a heroine of romance. Indeed, this passage prepares us for her insistence that she is entitled to promote the success of her first venture into political economy by doing 'what a man of business would do' and travelling to London on her own to visit publishers in person, since she has already been forced by deafness to become someone who must 'take [her] case into [her] own hands'.[42]

By this time, Martineau's talents as an author have already been recognised by her family, and the collapse of their finances – what Mary Jean Corbett calls her 'fortunate fall' from gentility – has freed her from a stultifying life of idleness.[43] It is her deafness, however, which saves her from any pressure to pursue more conventionally feminine means of self-support ('my deafness precluding both music and governessing') and which, more subtly, creates in her the independence of spirit which enables her, in this account, first to conceive and then to bring to fruition her plan for the *Illustrations of Political Economy* series, a literary venture so innovative that, at first, no one else can see its merits.[44] Martineau's account of her first struggle for success is one of heroic self-mastery and self-reliance, in which she 'stifled [her] sighs, and swallowed [her] tears, and wrote to one publisher after another' before going to the city to plead in person, tramping through the London streets from office to office. Although this is ostensibly an account of suffering, it also pulses with the anticipation of coming triumph: even as she is forced to lean against railings at the roadside, 'too giddy to stand without support', she is 'saying to myself, as I stood with closed eyes, "My book will do yet."'.[45]

Compared to the passages chronicling her abject childhood misery, the struggle to persuade anyone to publish her book is recounted almost with relish. It is as if Martineau feels most herself when pitted against an unsympathetic world, struggling for the recognition which is eventually wrested from unwilling but ultimately admiring onlookers. Her loneliness in London seems to be a part of her triumphant emergence as an authoress; her mother's summons that she should return home, at a time when she wants to stay and support herself through 'proof-correcting, and other literary drudgery', is met with 'disappointment, – I might almost say, horror'.[46] As Peterson points out, 'in Martineau's account, the domestic sphere becomes the site of pain, not a refuge from it';[47] given the pleasure Martineau takes in setting up her own household later on and providing herself with the domestic life without which, she insists, 'no true woman [. . .] *can* be happy', I would suggest that this is because 'the domestic sphere' at this point involves living with other people to whose wishes she must defer.[48] Martineau's domestic ideal, as we shall see, involves not only enjoying the position of head of the household but interacting with others only on strictly defined and limited terms. Her life in London is, even in its difficult beginnings, depicted as conducive to her happiness because it enables her to spend so much time alone, and to define herself above all as a solitary, self-supporting individual.

Martineau concludes the third section of her *Autobiography*, in which she has achieved her great literary breakthrough and established herself independently in London, with an explicit announcement that the miserable part of her life is now over:

> And now the summer of my life was bursting forth without any interval of spring. My life began with winter, burst suddenly into summer [. . .] I had now, by thirty years of age, ascertained my career, found occupation, and achieved independence . . .[49]

There is no doubt, then, that Martineau wants us to understand her as having triumphed over adversity; equally clearly, the 'great calamity' of deafness is constructed as one aspect of that adversity.[50] However, it is also the means by which she is able to triumph over the true threat to her 'career', 'occupation', and 'independence': namely, others' expectations, opinions, and demands. In her life in London, as we shall see, sociability continues to pose a threat to the independence on which her happiness is built, a threat which is managed through her deafness.

'A Grievous Waste of Time': The Pleasures of Work and the Perils of Sociability in London and North America

In the seven years between leaving Norwich in 1832 and travelling to North America in 1839, Martineau lived the life of a literary celebrity in London, apparently invited everywhere and enjoying a wide circle of eminent acquaintance. In the *Autobiography*, she depicts this period of (relative) sociability in a profoundly ambivalent way. It clearly gives her pleasure to describe the many famous people she knew; while it was a convention of autobiographies in this period that pen portraits be offered of the great and the good, the hundred pages she devotes to sketches of well-known friends, enemies, and acquaintances suggests that Martineau enjoyed dwelling on her society days. She explicitly contrasts the pleasures of social life in London to the pains of social intercourse in Norwich, where '[m]y deafness was terribly in the way, both because it made me shy, and because underbred people [. . .] are awkward in such a case', whereas in London:

> People began with me as a deaf person; and there was little more awkwardness about hearing, when they had once reconciled themselves to my trumpet. [. . .] I now found what the real pleasures of social intercourse are, and was deeply sensible of its benefits: but it really does not appear to me that I was intoxicated with the pleasure, or that I over-rated the benefit. [. . .] I always preferred my work to this sort of play. [. . .] [W]hile vividly remembering the seven years from 1832 to 1839, and feeling as gratefully and complacently as ever the kindness and attachment of friends, and the good-will of a multitude of acquaintances, I had no inclination to return to literary life in London after my recovery at Tynemouth, and have for ten years rejoiced, without pause or doubt, in my seclusion and repose in my quiet valley.[51]

As the latter part of this passage suggests, Martineau's praise of London society is heavily qualified by the insistence that she has now made herself completely independent of the need for such recognition and attention, and by some disparagement of its pleasures, even at the time: 'I always preferred my work to this sort of play'. Although she details her strenuous working habits, long days, and enormous productivity in such a way that we might suspect she is casting herself as a kind of martyr to her work, Martineau also stresses the pleasures of writing, which she calls 'the fulfilment of a natural function', making it sound almost effortless, in strong contrast to her depictions of social interaction'.[52] The greatest dangers to this pleasurable process of creativity and composition are, in Martineau's account, those posed by society.

Before launching into sketches of her acquaintance in London, she reprints a lengthy essay on the evils of 'Literary Lionism', essentially an extended complaint about the false social relations into which literary celebrities are led by their admirers. Not only does this entail 'a grievous waste of [. . .] precious time', but she casts others' praise as a threat to the quality of their work: 'Having gained much applause by a particular quality of his writing, he will be always trying to get more applause by a stronger exhibition of the quality, till it grows into pure extravagance'.[53] The only way of being 'safe', Martineau claims, is to be able to 'do without praise' – effectively, to possess the self-sufficiency which Martineau's experience of deafness has forced her to develop. Not only has she already faced up to the possibility of others' dislike and learned to do without their approval, but her deafness begins to appear in the text not as an unwelcome barrier to social interaction but as a kind of armour. When she is weathering the storm of controversy over her pro-Malthusian writing, she calmly replies to a 'literary lady' who warns her that insult may await an authoress who tackles such a subject: '[i]f any others should come whispering to me what I need not listen to, I shall shift my trumpet, and take up my knitting'.[54] As the *Autobiography* goes on, it is not simply that Martineau claims she is not wounded by adverse reaction to her work but that – in stark contrast to dangerous praise – others' disapproval is treated as salutary: 'my first trial in the shape of hostile reviewing was over, and I stood unharmed, and somewhat enlightened and strengthened'.[55] Sometimes, she goes further and suggests that she thrives on disparagement in the press; in a quoted diary entry of 1840, she observes that

> [i]t always is the way with me, and has been since I grew up, that personal mortifications [. . .] put me in a happy state of mind [. . .] It is always so with hostile reviewers; – the more brutal, the more animating, in a very little while.

Then, looking back at the *Quarterly*'s attack in 1833, she happily reflects, 'the reaction was immediate, and the rebound from mortification to joyful acquiescence was one of the most delightful experiences I have ever had'.[56]

Unlike hostile reviewers, well-meaning members of her family continue to be cast largely as irritants and distractions, from the brother who makes her late to the Queen's Coronation to the mother who – more seriously – undermines her health with her unreasonable emotional demands.[57] Martineau admits that she went to North America largely to escape her mother's company but also to evade the perils of London society, 'to break through [. . .] any love of ease and indulgence that might have arisen out of success, flattery, or the devoted kindness of my friends'.[58] Certainly, she does not have to contend with excessive 'flattery' or 'kindness' in North

America, where her outspoken support for Abolition soon makes her a target for derision in the press and threatens her personal safety – but a trip which the reader might expect to be nightmarish is in fact cast as profoundly restorative. Disgusted by the timidity of her hosts in Boston, whose fear of offending she casts as craven, Martineau seems to enjoy her notoriety. Of course, there is a political dimension to this relish, with Martineau's satisfaction at finding 'that in the South I am still reviled, as I was twenty years ago' reflecting her conviction that the cause of Abolition is one for which everyone should be willing to make sacrifices, but the sheer zest of her account makes it difficult not to conclude that the opposition she met with gratified her, at least to some extent, for its own sake.[59] This typifies the way that the Martineau of the *Autobiography* forms her identity in opposition to others' views and demands, so that she is most powerfully herself when alienated from those around her. Apparently much happier receiving death threats than praise which might turn her head, it is as if Martineau revels in the contrast between those who fear for her safety and suffer 'subservience to others' opinion' and her own indifference, because it so clearly marks her own development from the girl who was so miserably 'dependent [. . .] upon the opinion of others' and her present self as a woman who has '[o]n five occasions [. . .] found myself obliged to write and publish what I entirely believed would be ruinous to my reputation and prosperity' and emerged unscathed every time.[60]

By this point in the *Autobiography*, her deafness has ceased to function as a source of suffering or of timidity. The ultimate sign of her surmounting what she calls her 'false shame' is her adoption of an ear trumpet at the age of twenty-eight.[61] Whilst mentioning her deafness relatively rarely, she often refers to her ear trumpet, perhaps because she wanted to encourage its use in others, perhaps because, as Jennifer Esmail points out, it had become almost a symbol of her identity in the minds of her readers.[62] From the second half of the first volume onwards, she characterises (and judges) friends and acquaintances by their willingness to use her trumpet or to adopt one themselves: both Bulwer-Lytton and the Duke of Wellington are disparaged for their failure to use one despite their deafness, and a clear contrast is drawn between those friends who look upon her trumpet with 'shy dislike' and shout instead, and those whose 'good things were so dropped as to fall into my trumpet, without any trouble or ostentation'.[63] Rather than worrying about whether she is sufficiently adapting herself to their needs (as she urges her deaf reader in the 'Letter'), Martineau in fact comes to judge others for how well they adapt themselves to hers. Instead of trying to blend in, she seems to embrace the way that her trumpet marks her out; when she meets Charlotte Brontë, she does not instantly know her for Currer Bell, whereas Brontë knows her at once.[64] In North America, this marker adds to her personal danger and makes it harder for her to travel by river, when she learns that the anti-abolitionists 'had set a watch upon the steamboats, where I should be recognised by my trumpet'.[65] By this time, however, we feel that it would be so out of character for Martineau to go anywhere incognito that the trumpet seems only to be making manifest something that is profoundly true of her character. Her deafness has become an integral part of her identity, which is constructed through the isolation from others upon which both her work and her well-being depend. That isolation, of course, must never become total: sharing her opinions with the public is at the heart of her work, and communication through writing as much a duty ('Things were pressing to be said; and there was more or less evidence that I was the person to say them') as a need ('I wrote because I could not help it').[66] But praise, appreciation, and even congenial company are depicted as dangerous pleasures, to be enjoyed in strict moderation – whereas opposition is on every occasion shown as conducive to the independence of spirit by which Martineau increasingly defines herself.

Conclusion: Domestic Bliss and Splendid Isolation in the Lakes

In the first edition of Martineau's *Autobiography*, the frontispiece illustration for the first volume shows a young Martineau with an amiably interested, perhaps listening expression, her hair swept away from her face to expose her ears, one of which is turned towards the viewer and cupped by her hand. It is, as Graeme Gooday and Karen Sayer point out, a gesture strongly associated in the period with deafness, and marks Martineau unmistakably as someone who has experienced hearing loss.[67] The illustration of the second volume, by contrast, shows a middle-aged Martineau wearing a lacy cap, her ears obscured: no observer who did not already know of Martineau's deafness would infer it from this image. Although the variation in the illustrations may be simply the result of the images available to the publishers, it neatly expresses the shift that takes place between the first two-thirds of Martineau's narrative and what she calls her 'mild and sunny' Autumn, in which she has been able to 'relish life, without drawback'.[68] In this last chapter of her life story, Martineau's deafness recedes from the narrative almost completely, to the point that it is mentioned in passing only a handful of times, and then only in unsatisfactory social situations.[69] It is as if once Martineau has achieved a perfect relationship with sociability – one which is rooted in deliberate self-isolation but still enlivened by the pleasures of sought-after controversy and carefully managed social interactions of her choosing – her deafness ceases to matter so completely that it effectively ceases to exist, at least as the 'privation' and 'disability' of the 'Letter'.[70] In this final stage of her life, Martineau apparently achieves the total control over her social interactions which she has always sought and is able to find a perfect balance between absolute solitude and wished-for recognition, in which the dangers of sociability can be successfully managed and the difficulties of deafness accordingly neutralised to the point that they no longer register in the text at all.

Martineau's life in the Lake District follows the five years she spent as an invalid in Tynemouth, which in her account were the result of the 'excessive anxiety of mind' caused by the unwelcome and enforced sociability of her life in London, which she was obliged to share with uncongenial relations.[71] Although she embraces the solitude of her sickroom eagerly at first, finally able to live alone and feeling 'comparatively happy in my release from responsibility, anxiety, and suspense', her life at Tynemouth represented an opposite extreme of isolation and immobility which was painful to remember: she prefers, she says, to 'proceed as if that awful chasm had never opened in my path'.[72] After experiencing through mesmerism a complete cure which left onlookers baffled (and in some cases highly indignant), Martineau suggests that she has found an ideal balance between the two states. By choosing not to resume her literary life in London but instead to build a house near Ambleside in the Lake District, she is able to avoid a return to life in society but also to entertain whenever she wishes.

Martineau draws the reader's attention, repeatedly, to the comparative solitude of her Lake District life, explaining that she decided not 'to visit generally in Ambleside', 'declined the dinner and evening engagements offered', avoided entertaining on the grounds that '[i]t did not suit me to give parties', and escaped unwanted visitors by going away in the busy summer months.[73] Social intercourse still occurs, of course – maidservants are variously befriended, mesmerised, and married off, visits to selected friends are paid, and favoured nieces are entertained from time to time – but these interactions all take place strictly on Martineau's terms, and their satisfactions seem to be outweighed by what are cast as the surpassing pleasures of solitude. Whilst insisting that 'few [. . .] are of a more radically social nature than myself', Martineau claims that

> there has never been, since I had a home of my own, an evening spent in the most charming intercourse that I would not have exchanged (as far as the mere pleasure was concerned) for one of my ordinary evenings under the lamp within, and the lights of heaven without.[74]

In her account, this withdrawal from rural society is partly justified by the offence her opinions would inevitably give her neighbours at some point; as though to guarantee the truth of this, she published in 1851 *Letters on the Laws of Man's Nature and Development*, which made public her deconversion from Unitarianism and avowal of atheism and Necessarianism. Although Martineau reports herself happily surprised by the book's warm reception, she nonetheless details all the small snubs – from the 'diminution of letters from mere acquaintances' to being blackballed by the Ambleside book club (which she had never wanted to join) – with what can only be described as relish.[75] Shalyn Claggett suggests that this 'overdetermined' reaction to a relatively mild backlash is a part of Martineau's confessional mode:

> She performs confession not to be accepted by the social order [. . .] but to announce her willingness to be separated from it, like a martyr willing to be, as she puts it, 'burned at the stake' (*Autobiography*, II, p. 346).[76]

I see Martineau's decision first to publish an account of her deconversion and then gleefully to detail the hostile reaction which precedes praise and acceptance as consistent with her desired relation to others – from members of her own family to the wider, personally unknown public – throughout the *Autobiography*. From the onset of her deafness, Martineau defines herself as alienated from others, forced to reject advice and go her own way, knowing her own power through resistance to others' opinions and indifference to their disapproval, and then, through this brave perseverance, finally winning the approval she has proved she does not need. Her account of the publication of the *Letters* perfectly fits this pattern and forms a kind of apotheosis to her narrative.

Right at the end, however, a different note is struck. After learning that she has fatal heart disease in 1855, Martineau invites her 'beloved niece' to come and stay with her permanently and enjoys her 'incomparable nursing'.[77] In this final episode, Martineau happily reflects that '[n]ever was a person more nobly "friended", as the Scotch have it' – indeed, her only anxiety in what she believes to be her final months is the fear, reiterated several times, that 'after all the discipline of my life, I should be spoiled at the end of it'.[78] In fact, the doctor's death sentence proved premature: Martineau lived another twenty years, so she had ample time to enjoy this new and newly unthreatening relationship with her own work, her own senses, and above all with those around her. In a moment of candour, she revises and complicates her earlier claim to be of a 'radically social nature' by admitting, 'I have hitherto loved solitude perhaps unduly; partly, no doubt, on account of my deafness'.[79] Now, however, free from responsibility, from work, and from the wearisome obligation to 'husband my remaining hearing', Martineau feels entitled to take 'a free and lavish holiday before I go', in which she can 'relish, very keenly, the tending of affection'.[80] It is a touching ending to an autobiography which has so powerfully expressed the dangers of sociability and of dependence on others. Believing herself to be dying, Martineau is finally able to enjoy the 'affectionate intercourses' which seemed previously to threaten her sense of self. Although Martineau never uses the word, what she seems to discover in these final pages are the possibilities and pleasures of care.

Notes

1 Harriet Martineau, 'Letter to the Deaf', *Tait's Edinburgh Magazine* (1834), 174–9 (p. 174).
2 Ibid., pp. 175–7.
3 Ibid.
4 Ibid., pp. 176, 179.
5 Ibid., pp. 174–6; Harriet Martineau, *Autobiography*, 2 vols (London: Virago, 1983 [1877]), I, p. 76.
6 Martineau, *Autobiography*, II, p. 324.
7 Martineau, 'Letter', pp. 174, 175.
8 Martha Stoddard Holmes, *Fictions of Affliction: Physical Disability in Victorian Culture* (Ann Arbor: University of Michigan Press, 2004), pp. 153–4.
9 Martineau, 'Letter', pp. 176, 177.
10 Ibid., p. 179.
11 In, for example, Martineau, *Illustrations of Political Economy* (1832–34), *Letters on the Laws of Man's Nature and Development* (1851), and *The Hour and the Man* (1840).
12 See, for example, 'Miss Martineau's Monthly Novels', *Quarterly Review*, 49 (1833), 136–52.
13 Dorothy Mermin, *Godiva's Ride: Women of Letters in England, 1830–1880* (Bloomington: Indiana University Press, 1993), p. 100.
14 Linda H. Peterson, *Victorian Autobiography: The Tradition of Self-Interpretation* (New Haven: Yale University Press, 1986), pp. 124, 143.
15 Peterson, *Becoming a Woman of Letters: Victorian Myths of Authorship, Facts of the Market* (Princeton, NJ: Princeton University Press, 2009), pp. 74–5; Susan F. Bohrer, 'Harriet Martineau: Gender, Disability and Liability', *Nineteenth-Century Contexts*, 25.1 (2003), 21–37 (pp. 23–4).
16 Martineau, 'Letter', p. 126.
17 Ibid., p. 129.
18 I am aware that those who are culturally 'Deaf' may reject altogether the alignment of deafness and disability; as Lennard J. Davis explains, 'the Deaf do not regard their absence of hearing as a disability, any more than a Spanish-speaking person would regard the inability to speak English as a disability' (*Enforcing Normalcy: Disability, Deafness, and the Body* (London: Verso, 1995), p. xiv). However, I agree with Jennifer Esmail that Martineau 'would have personally identified as deaf rather than Deaf' (*Reading Victorian Deafness: Signs and Sounds in Victorian Literature and Culture* (Athens: Ohio University Press, 2013), p. 110.); and since she understood her deafness as a disability, that is the dominant framework of this chapter.
19 George Eliot, *The Mill on the Floss* (Oxford: Oxford University Press, 2008 [1860]), p. 522; Charles Dickens, *Our Mutual Friend* (Oxford: Oxford

University Press, 2008 [1864–65]), p. 732; Wilkie Collins, *The Moonstone* (Oxford: Oxford University Press, 2008 [1868]), p. 23; Harriet Martineau, *Deerbrook* (London: Virago, 1983 [1839]), p. 36.
20 Talia Schaffer, *Romance's Rival: Familiar Marriage in Victorian Fiction* (New York: Oxford University Press, 2016), p. 160 and passim.
21 Harriet Martineau, *Life in the Sick-Room*, ed. Maria H. Frawley (Peterborough, ON: Broadview, 2003 [1844]), p. 117.
22 Davis, *Enforcing Normalcy*, p. 83. For a more detailed account of the rise and significance of the oralist movement, see Esmail, *Reading Victorian Deafness*.
23 Esmail, p. 110.
24 Martineau, 'Letter', p. 174; Martineau, *Autobiography*, I, p. 74.
25 Martineau, *Autobiography*, I, p. 18.
26 Ibid., p. 19.
27 Ibid., p. 10.
28 Ibid., pp. 10–1.
29 Ibid., p. 10.
30 Ibid., p. 45.
31 Ibid., p. 150.
32 Mermin, *Godiva's Ride*, p. 102.
33 Martineau, *Autobiography*, I, p. 48.
34 Ibid., p. 46.
35 Ibid., p. 48.
36 Ibid.
37 Ibid., pp. 17, 74.
38 Ibid., p. 77.
39 Ibid., p. 78.
40 Martineau, 'Letter', p. 177.
41 Martineau, *Autobiography*, I, p. 76.
42 Ibid., p. 164.
43 Ibid., pp. 120, 141; Mary Jean Corbett, *Representing Femininity: Middle-Class Subjectivity in Victorian and Edwardian Women's Autobiographies* (New York: Oxford University Press, 1992), p. 90.
44 Martineau, *Autobiography*, I, p. 143.
45 Ibid., pp. 163, 170.
46 Ibid., pp. 148–9.
47 Linda H. Peterson, *Traditions of Victorian Women's Autobiography: The Poetics and Politics of Life Writing* (Charlottesville: University Press of Virginia, 1999), pp. 61–2.
48 Martineau, *Autobiography*, II, p. 226.
49 Martineau, *Autobiography*, I, p. 180.
50 Ibid., p. 70.
51 Martineau, *Autobiography*, I, pp. 185–6.
52 Ibid., pp. 187–8.
53 Ibid., pp. 284, 287.
54 Ibid., pp. 201–2.
55 Ibid., p. 206.
56 Martineau, *Autobiography*, II, pp. 158–9.

57 Ibid., pp. 121, 150–1.
58 Ibid., p. 3.
59 Ibid., p. 40.
60 Martineau, *Autobiography*, II, p. 39; Martineau, *Autobiography*, I, pp. 70, 199.
61 Martineau, *Autobiography*, I, p. 125.
62 Ibid., p. 125; Esmail, *Reading Victorian Deafness*, p. 170.
63 Martineau, *Autobiography*, I, pp. 353–4, 372, 323; Martineau, *Autobiography*, II, p. 338.
64 Martineau, *Autobiography*, II, p. 326.
65 Ibid., p. 48.
66 Martineau, *Autobiography*, I, p. 188; Martineau, *Autobiography*, II, p. 95.
67 See Graeme Gooday and Karen Sayer, 'Introduction', in *Managing the Experience of Hearing Loss in Britain, 1830–1930* (London: Palgrave Macmillan, 2017), pp. 1–16.
68 Martineau, *Autobiography*, I, p. 180; Martineau, *Autobiography*, II, p. 205.
69 She mentions her deafness, for example, when explaining why she does not often want to visit the excessively sociable William Wordsworth, who provides a cautionary example of the author who is too dependent on others' praise and interest; it is also mentioned when she has uncongenial dinner guests, and as a possible (though surmounted) hindrance to her giving a series of public lectures (Martineau, *Autobiography*, II, pp. 237, 395, 305).
70 Martineau, 'Letter', p. 179.
71 Martineau, *Autobiography*, I, p. 146.
72 Martineau, *Autobiography*, II, pp. 146, 213.
73 Ibid., pp. 221, 266.
74 Ibid., p. 417.
75 Ibid., pp. 357, 304.
76 Shalyn Claggett, 'Harriet Martineau's Material Rebirth', *Victorian Literature and Culture*, 38.1 (2010), 53–73 (p. 67).
77 Martineau, *Autobiography*, II, p. 433.
78 Ibid.
79 Ibid., p. 446.
80 Ibid.

3 Virginia Woolf's *Mrs Dalloway* and the Biopolitics of Interwar Noise Abatement

Anna Snaith

I am writing in lockdown during the 2020 COVID-19 pandemic, and what would be routine processes of thinking and expressing seem utterly disorientating. The stasis and sameness of the quotidian jar with global connectedness through crisis and shared loss. The staggering global death toll and tales of the medical front line rupture the reductions of this new existence. With these constraints has come not only an altered soundscape but differently attentive listening. In the newly hi-fi soundscape, discrete sounds emerge with particular clarity: cries of children, a car door slamming. The rare overhead buzz of a plane demands auditory attention, and the ubiquitous sound of sirens is a sonic reminder of the distress within hospitals and places of care. Amidst the reduction in noise print, we are experiencing what R. Murray Schafer called 'ear cleaning': retuning ourselves to the sounds of the natural world.[1] As is so often remarked during times of conflict and crisis, birdsong is seemingly amplified, listened out for as a sonic reminder of 'normal' time with its predictable signals. Sonic events, such as the Thursday evening 'Clap for Carers' in the UK, have been crucial for both solidarity and expression. Communities have found their own music-making routines, whether charivari or classical street concerts. Projects like #StayHomeSounds or the Silent Cities Project – 'a participatory monitoring project of an exceptional modification of urban soundscapes' – have been developed or expanded to capture the sonic effects of the pandemic.[2] With quiet and silence under increased scrutiny comes an altered understanding of noise, as associated with a pre-pandemic life of speed and consumption.

The first half of the twentieth century was also a period of shifting sonic environments and the discourses used to describe them. From the rise in industrial lo-fi hum, to the extreme noise environments occasioned by the two World Wars, this period played earwitness to rapid variations in urban soundscapes. The introduction of sirens on emergency vehicles and an increase in loudspeakers and amplified sound brought new pitch and volume to urban ears, just as the rise in air travel meant surround-sound, mechanical noise. While sound recording and telephonic communication had been

DOI: 10.4324/9780429291616-4

in existence for decades, the popularisation of these technologies, alongside the inauguration of the BBC in 1922, had a transformative impact on sound levels and where and how listeners experienced them. Modern auditors were practised in aural streaming or filtering out the din of the machine.

Such changes brought about a preoccupation with mechanical sound, evoked so powerfully in A. H. Davis's *Noise* (1937): 'the irregular rattle of machinery, the shriek of brakes, the crescendo of violently accelerating motor-cycles, the shattering explosions of road-breaking drills, burst rudely upon our consciousness and stir our resentment and fears.'[3] Specific keynote sounds became auditory obsessions, such as the motor horn, described in hyperbolic terms by medic Dan McKenzie in *City of Din* (1916): 'never before in the whole raucous history of din have such fiendish contraptions split the air.'[4] Ironically, the horn's increased usage was due to noise reduction in motor vehicles and the dangers resulting from growing motor car ownership in an environment at various points without speed limits, drivers' licences or traffic lights. Noise offers an alternative 'earspective' on the city: one often muted by emphasis on the changing visual paradigms of the modernist period.

Writing at mid-century, Aldous Huxley designated his period the 'Age of Noise', and this chapter will tune in to the complex and often conflicting narratives and discourses through which noise was understood.[5] More specifically, I argue that Virginia Woolf's *Mrs Dalloway* (1925) represents and recalibrates the interwar noise debates which emerged, as does the novel, out of the cacophony of the First World War. The novel is part of a period obsession that found later and more concrete articulation in the noise abatement work of the Anti-Noise League (later Noise Abatement League) founded by Thomas Horder in 1933. Woolf's critique of eugenicist thinking and the treatment of those (whether women or war veterans) experiencing mental ill-health anticipates the harnessing of such thinking in the context of the noise debates. The paradigms through which urban, particularly mechanical, sound was understood were generated in relation to biopolitical discourses of efficiency and regulation: exactly those discourses that Woolf exposes so forcefully in relation to systems of power and their creation of value. In the interwar period, 'unwanted' sound was associated with waste and excess, whether excessive volume, the background hum of the machine or 'noisy', aberrant bodies.

Given that the designation of sound as 'noise' relies on context, or the capacity of sound to distract and disrupt in some environments and not others, this makes narratives about noise all the more illuminating in relation to the preoccupations of a particular age. Hearing, far from its recurrent perception as a passive sense, is an amalgam of the biological, cultural and political. The physiology and phenomenology of hearing braid with technological and historical determinants to create individual and collective sonic meaning-making. These narratives find expression, confirmation and resistance in the pages of literary fiction. This happens at the level of soundscapes, in the way language bears and anticipates the rhythms and sounds of speech, and in the interplay between these aspects of content and form. Literary modernism, in particular, has often been described in relation to a crisis or confusion of the senses: its synaesthetic motivations undergird a multimedia impulse that stretches the capacity of the word and the text. Modernist writers expanded and experimented with the ways a seemingly silent literary text might represent sound, whether, for example, in the inclusion of staves of music, in the jazz novel or in the representation of industrial soundscapes. Douglas Kahn employs the category of '*all sound*' to describe the acoustics of literary modernism: like a microphone or phonographic recording, the text tunes in to a full range of acoustic signals, and the modern reader has to filter and decide what to prioritise.[6] Techniques of the modernist novel, such as stream-of-consciousness writing and an interest in representing internal thought, might be read in tandem with the technological capacity for amplification or auscultation, or, as Angela Frattarola has argued, with the introduction of headphones.[7]

Many modernist writers were early noise theorists, concerned with the anarchic and aesthetic possibilities of noise: its complex frequencies a metaphor for modernity itself. The cacophony of urban modernity provided creative fuel as modernists sought to capture and represent an ever-widening array of sounds, those amplified and recorded by the increasingly popular sound technologies. Writers not only opened their ears to the noises of the modern but were drawn conceptually to the immersive and revolutionary possibilities of noise. We need only think of the Vorticists' magazine, *BLAST*, and of more general typographical experiments that evoke high-decibel utterances through use of colour font, or size.

But much of the scholarship on modernism and noise has tended to focus on the avant garde, in particular Italian Futurist Luigi Russolo's manifesto *The Art of Noises* (1913) and the instruments – *intonarumori* – that he built and performed. In considering Virginia Woolf's engagement with noise, we find not only a pacifist alternative to the celebration of war noise found in the writing of Russolo and F. T. Marinetti but also a rather different perspective on the connective and aesthetic possibilities of literary sound. Woolf's oeuvre demonstrates a preoccupation with the auditory, from her representations of cacophonous urban environments to her call to 'listen in to the past' articulated in 'A Sketch of the Past' (1939).[8] In 'Character in Fiction' (1924), Woolf describes the cataclysmic transformation of the Edwardian literary scene as an acoustic event:

> And so the smashing and crashing began. Thus it is that we hear all round us, in poems and novels and biographies [. . .] the sound of breaking and falling, crashing and destruction. It is the prevailing sound of the Georgian age.[9]

The demolition of the conventions of grammar and novelistic structure by the likes of James Joyce or Dorothy Richardson, she writes, is a 'vigorous and stimulating sound in my ears.'[10] In her late essay, 'Thoughts on Peace in an Air Raid' (1940), the hypersensitive hearing of the air raid sets off a sonic chain not of echoes but of dialectical sounds. Lying in the darkness alert to the sonic danger signs, the

> drone of the planes is now like the sawing of a branch overhead [. . .] Another sound begins sawing its way in the brain. 'Women of ability' – it was Lady Astor speaking in *The Times* this morning – 'are held down because of a subconscious Hitlerism in the hearts of men.'[11]

The essay foregrounds not only the auditory elements of wartime but the ways in which the external soundscape resonates with internal sonicity, thereby blending the phono*graphic* and articulated speech (here rendered in speech marks). Woolf's thinking on noise, found sound and their literary representation is multi-layered in its movement across the socio-cultural, technological and aesthetic: across content, context and form.

While early scholarship on Woolf and sound tended to focus on art music, an ever-increasing body of criticism is now addressing Woolf's multi-layered interest in all manner of sonic phenomena, from noise to rhythm to sound technologies (particularly radio).[12] In response to Douglas Kahn's assertion that modernism has been seen but not heard, a wealth of recent publications have produced something of an acoustic turn in modernist studies.[13] Taken in tandem with work on the soundscapes of urban modernity by James Mansell, Karin Bijsterveld and Emily Thompson, a more richly historicised and theorised understanding of modernism's sonic contours is beginning to form.[14] That said, much of the recent work in sound studies bypasses the literary as a site of formation and cultural understanding. As Mansell reminds us, 'writers and writing play a role not only in capturing and commenting upon our cultures of hearing, but also in shaping them.'[15]

In addition, there is still relatively little attention in literary sound studies to one of Woolf's 'noisiest' novels. *Mrs Dalloway* registers with precision and nuance the soundscape of central London while echoing with the sonic legacies of the First World War. Woolf wrote the novel while experiencing a dramatic shift in her sonic environment as she moved, in 1924, back to London from Richmond, where she had been living since 1914. She deliberates, in her diary, on habituation to urban cacophony and retraining herself aurally for life in the metropolis:

> the habit of living at 52 Tavistock Sqre is not quite formed, but doing well. Already I have spent a week without being bothered by noise. One ceases to hear or to see. [. . .] I notice things much less than I did 10 days ago. Soon I shall be making a habit of life in this room.[16]

She monitors her own sensitivity to noise and, appropriately, ends up literally writing and reading *Romeo and Juliet* at the open window, an apt metaphor for the manner in which her work has its ear to the city, open to its noises. This cluster of diary meditations becomes a riff on acclimatisation to the onslaught of urban phenomena: in particular on how the overwhelmed ear adapts to sonic complexity. They show her welcoming the jolt of sonic disruption; as she was to write later: 'the shock-receiving capacity is what makes me a writer.'[17] Woolf's openness to auditory disturbance offers an alternative paradigm in relation to modern shock which is so often conceived as tactile or visual.

More broadly, *Mrs Dalloway* emerges out of a post-war preoccupation with noise. Practitioners working across a range of disciplines like the creative arts, medicine, acoustics and experimental psychology attempted to define, measure, interrogate, listen in to and legislate against noise. In the avalanche of books on noise published in the 1920s and '30s, noise was variously defined as 'unwanted' or 'anarchical' sound or via adjectives such as 'nonperiodic', 'irregular' or 'unbalanced'. In a period characterised by political and economic crisis, that uncertainty had an aural register, but scholars have been slower to 'hear' the sonic preoccupations of interwar Britain. Noise abatement work grew apace in the late 1920s; many writers of the period such as Aldous Huxley, H. G. Wells, Rebecca West and J. B. Priestley were involved with the Noise Abatement League (NAL) and wrote for its magazine: *Quiet*. In its pages, noise is discussed as a real acoustic phenomenon but is also the focal point for a number of other anxieties that have to do more generally with human control of technology and the impact of urban life on the human body and its nervous system. Noise became shorthand for the perceived dangers of an increasingly mechanised and technological modern world. As Dan McKenzie wrote in *City of Din*: 'civilisation is noise. At least modern civilisation is. And the more it progresses the noisier it becomes.'[18]

This interest in the material impact of invisible sound waves led to calls for the regulation and reduction of the din. The project of 'taming the man-made acoustical medley' is a 'vast enterprise' that requires the 'co-operation of the whole nation', urged Norman McLachlan in yet another book called *Noise* (1935).[19] In earlier decades and centuries, narratives of acoustic sensitivity had been distinctly class bound, but in the interwar period this ostensibly gave way to a broader concern about the detrimental effects of everyday urban sound on the national and individual body. In effect, noise became a public health issue.

It is no coincidence that this was the period in which it was possible to measure noise with a greater degree of objectivity thanks to a unit (decibel) and instruments (noise meters) that did not rely on the operator's hearing. Although the new audiometers were extremely unwieldy to move and operate, they were immediately put to use for the creation of sound maps and charts by those engaged in noise abatement work. While, in 1886, Thomas Barr had conducted the first experiments into the effects of industrial noise on Glaswegian boilermakers, it was not until the 1920s and '30s that the effect of noise on the body became a subject of sustained scientific enquiry.[20] The government's Industrial Health Research Board (founded initially in 1918 as the Industrial Fatigue Research Board) conducted a number of noise experiments into the impact of noise in a range of workplaces. These included measuring the effects of noiseless typewriters on worker output and investigating the 'nervous temperament' in a range of noise environments.[21] In 1932 and 1935, the IHRB commissioned studies by H. C. Weston and S. Adams into the effects of high-decibel industrial noise on weavers in a Lancashire weaving shed where sound intensity levels measured 96 dB. Despite the IHRB's attempts to distance itself from F. W. Taylor's scientific management and its emphasis on profit maximisation, these experiments rarely tested the impact (psychological or physical) on workers' hearing, instead recording the impact of sound on productivity levels. Noise became cause and symbol of the 'inefficient' worker.

Virginia Woolf's *Mrs Dalloway* is as interested in those citizens deemed 'inefficient' as it is in the sonic diversity of London. These concerns, I argue, come together in competing narratives about noise and its impact. The novel registers a post-war preoccupation with urban sound that offers anticipatory resistance to public, state-sponsored narratives about noise that emerged later in the interwar period. The novel revels in urban noisescapes: the 'brass bands; barrel organs; in the triumph and the jingle and the strange high singing of some aeroplane overhead.'[22] Its characters walk the city streets with open ears, receptive to registering the cacophony of London, in part a post-war, post-pandemic response to urban movement and dynamism. The opening scene in which Clarissa exits her house thinking of internal doors being 'taken off their hinges' for the evening party contains sonically inflected memories of Bourton and her teenage years.[23] The motif of opening doors, as well as the sound of the word 'hinges', provides an opening into the past triggered by the 'little squeak of the hinges, which she could hear now.'[24] Clarissa inhabits her house and understands those people working within it through its soundscape:

> faint sounds rose in spirals up the well of the stairs; the swish of a mop; tapping; knocking; a loudness when the front door opened; a voice repeating a message in the basement; the chink of silver on a tray.[25]

Attuned to the sounds of labour and movement, Clarissa performs a kind of auditory surveillance over the servants preparing for her party. The novel tunes in to the role of hearing in navigating particular spaces in the present, as well as in triggering and accessing memories from the past.

A novel about shell shock is inevitably a novel about noise. Woolf registers the sonic legacies of the war and articulations of the material impacts of sound. The deafening surround sound of artillery and exploding shells was extreme (140–180 dB) and caused vibrations forceful enough not just to shock but to shatter ear drums and overwhelm the body's sense organs. War noise was often described as felt rather than heard, and even those on the home front registered the din. In her essay 'Heard on the Downs: The Genesis of Myth' (1916), Woolf describes 'this sinister sound of far-off beating' as the noise of war 'roll[s] across the bare uplands' of the South Downs.[26] Initial definitions of shell shock focused on the organic damage caused by the sound and force of exploding shells. Neurologist Frederick Mott used brain dissection to attempt to find noise-induced lesions on nerves, and war doctor Charles Myers, who coined the term 'shell shock' in 1915, wrote that 'high-frequency vibrations caused by an exploding shell might [. . .] conceivably produce an invisibly fine "molecular" commotion in the brain which, in turn, might produce dissociation.'[27] Medic Edwin Ash wrote of the impacts of war noise:

> the noise alone is sufficient to cause such vibrations in the brain and special sense organs as not infrequently to overwhelm them; shells and bombs, mines and aerial torpedoes, bursting with ceaseless din, have combined in such an assault of sound as the human nervous system has never before been asked to sustain.[28]

Although, very rapidly, psychological explanations of shell shock entered the frame, the war brought about a renewed interest in neurasthenia or nervous disorders caused by sound, and while the extreme noise levels of the conflict zone did, of course, cause physical injury (deafness, burst ear drums, tinnitus), the sonic legacy of war was a new understanding of everyday urban din as physically dangerous. Sound was understood as random, uncontrollable and visceral.

Woolf depicts the auditory environment of London by means of war noise. The text's paired protagonists, Clarissa and Septimus, never meet but have sonic as well as visual points of connection, such as the noise of an exploding tyre: 'a violent explosion which made Mrs Dalloway jump', 'oh! A pistol shot in the street outside!'[29] Urban hubbub is registered, even by civilians, via the sounds of war, and the narrative enacts bodily shock through exclamation and interruption. One of Woolf's diary entries provides a clue to the genesis of Clarissa's response:

> In St James Street there was a terrific explosion; people came running out of Clubs; stopped still & gazed about them. But there was no Zepplin or aeroplane – only, I suppose, a very large tyre burst. But it is really an

> instinct with me, & most people, I suppose, to turn any sudden noise, or dark object in the sky into an explosion, or a German aeroplane.[30]

The memory and anticipation of trauma registers aurally, to quote Paul Saint-Amour, as a 'proleptic mass traumatization: a pre-traumatic stress syndrome.'[31] The sound of the sky-writing aeroplane in *Mrs Dalloway* 'bore[s] ominously into the ears of the crowd', an image of materialised sound invading the communal body.[32]

Characters are yoked and bound together by the sound of bells. Peter Walsh hears the 'light high bell of the ambulance' come to attend to Septimus's suicide, and the chiming of Big Ben is a sonic refrain repeated four times in the novel.[33] Woolf's insistence on the need to hear rather than silence the horrifying noise of war can be found in her description of the 'direct, downright' chime as 'leaden circles' that 'dissolved in the air.'[34] The sound waves of the peals are materialised as shells which, although dissolving rather than exploding, act as a reminder of the conflict. That said, the bell's silencing during the second half of the war means that, in 1923, its chime is an acoustic reminder of peacetime and of the contextual nature of hearing. On 31 December 1923, the BBC began broadcasting the sound of Big Ben (rather than bells played in the recording studio), thus taking the sound beyond its immediate hearing range. Elizabeth Bowen's *To the North* (1932) is apposite here in terms of the remediation of Big Ben's chimes: 'When Big Ben struck, there was no one to whom to say, as she [Pauline] always did, that this made her think of the wireless.'[35] Woolf's oft-discussed motif of the bells is not only about temporal regularity and sonic regulation of a community in earshot but about the layered contingencies of hearing and sonic associations.

The novel also registers what Yaron Jean has called a 'wartime sonic-mindedness'.[36] The conditions of trench warfare required a hyper-alertness to sound, given the reduced vision occasioned by the terrain, the gas and the barbed wire. Survival depended on hearing. More specifically, it depended on the ability to identify the sounds of different types of explosives as well as to detect direction and spatial co-ordinates through sound. In a series of letters to Joseph Conrad – his 'notes upon sounds' written from France in September 1916 – Ford Madox Ford demonstrates this granular attention to war noise by itemising and visualising the distinct sounds of war.[37] In particular, he describes the auditory effects of artillery in different types of landscape: 'on dry down land the sound is much *sharper*; it hits *you* and shakes *you*. On clay land it shakes the ground & shakes you thro' the ground.'[38] Ford details at length the spatial and atmospheric effects, such as sounds seeming closer in wet weather, or the amplifying effects of woodland on the noise of heavy artillery.[39] This preoccupation with not only the sonic co-ordinates but also the volume and pitch of wartime noise indicates both an attempt to control or glean crucial information from sounds emanating from unseen and dangerous sources and a kind of muting of the threat to life signalled by the soundscape. Ford also experiments with onomatopoeic renderings:

> shells falling on a church: these make a huge '*corump*' sound, followed by a noise like crockery falling off a tray – as the roof tiles fall off. If the roof is not tiled you can hear the stained glass, sifting metallically until the next shell [. . .] screams of women penetrate all these sounds.[40]

Ford's phonography utilises a variety of modes to render a layered soundscape of destruction. The act of writing noise becomes another form of ordering or differentiation; listening is a sense to be trained and honed in the conflict zone.

Septimus exhibits this kind of auditory hyper-alertness and hyperacusis. His perception of his wife, Rezia's, crying is one such example: 'Far away he heard her sobbing; he heard it accurately, he noticed it distinctly; he compared it to a piston thumping. But he felt nothing.'[41] While shell shock has muted his emotional response, his sensory perception, including hearing, is on alert. Background noises, such as Rezia sewing, are amplified and analysed: 'when she sewed [. . .] she made a sound like a kettle on a hob; bubbling, murmuring.'[42] He experiences auditory hallucinations, such as the sparrows singing 'freshly and piercingly in Greek words', and the 'sound of water' filling his sitting room.[43] He 'was always hearing something new; listening with his hand up. But she [Rezia] heard nothing.'[44] His world is full of screams, the 'cries of people', 'voices of the dead'; a mouse squeaking or curtain rustling signals Evans's return.[45] His senses, disordered and confused, create synaesthetic impressions in which music becomes 'shocks of sound which rose in smooth columns (that music could be visible was a discovery)'.[46] Septimus's wartime acoustic trauma results in an amplified and oppressive soundscape, and he is constantly attempting to mitigate its overwhelming impact by listening in and analysing its component parts.

Mrs Dalloway registers the aftershocks of the cacophony of war, but it also offers a different kind of conceptualisation of the sonic. This is one that construes hearing, or the transfer of energy through vibration, as a mode of connection or resonance. Woolf was clearly intrigued by the metaphorical potential of the physics of sound and the physiology of hearing. Her writing abounds with descriptions of hearing that exemplify the physiological transfer of sound waves from a vibrating object to the ear drum and inner ear where fluid vibrations become electrical impulses. Septimus listens to his neighbour in the crowd sound out the letters of a skywriting aeroplane:

> 'K . . . R . . . ' said the nursemaid, and Septimus heard her say 'Kay Arr' close to his ear, deeply, softly, like a mellow organ, but with a roughness in her voice like a grasshopper's, which rasped his spine deliciously and sent running up into his brain waves of sound which, concussing, broke. A marvellous discovery indeed – that the human voice in certain atmospheric conditions [. . .] can quicken trees into life.[47]

With appropriate literary license, Woolf traces the passage of sound waves from vocal mechanism through to the brain. The reader not only hears the phonetic sounding of the letters but also the means by which the sonic message is transferred from one body to another. Woolf writes auditory process at the same time as reconfiguring the language of shell shock which associated sound with bodily injury, often in relation to 'concussion', a key word in wartime descriptions of organic injury from noise. Sound is physically invasive but in a pleasurable, spine-tingling way, energising rather than terrifying.

The phenomenon of sympathetic vibration runs through the novel, creating bonds between humans but also a wider sonic ecology that encompasses inanimate matter. Woolf makes reference to Clarissa's girlhood reading of John Tyndall, whose book *Sound* (1867) had popularised Hermann von Helmholtz's findings in acoustical physiology.[48] Von Helmholtz's *On the Sensations of Tone* (1863) had revolutionised the physiological understanding of how the ear hears complex tones by describing the workings of the ear in terms of sympathetic vibration or cochlear resonance. Hair-like receptors (corti) vibrate, he argued, in response to different frequencies, creating his idea of the 'nervous piano'. As critics such as Veit Erlmann and John Picker have shown, these discoveries (although not quite accurate) had immense cultural impact.[49] In *Mrs Dalloway*, sound creates connections between Septimus's body and the natural world in an acoustic ecology: 'leaves being connected by millions of fibres with his own body [. . .] Sounds made harmonies with premeditation; the spaces between them were as significant as the sounds.'[50] One could also argue that Woolf hangs the whole notion of paired protagonists who never meet around the idea of sympathetic vibration or resonance, from the tyre explosion to which they both respond to the *Cymbeline* quotation ('Fear no more the heat o' the sun') they both articulate to the ways in which Clarissa experiences Septimus's suicide, spiritually ('[s]omehow it was her disaster – her disgrace') as well as physically ('her body went through it').[51] Urban noise is the starting point for this series of echoes and connections.

The war and its decibel levels not only caused acoustic injury but produced discursive models of the body's finely tuned hearing apparatus as overwhelmed and depleted. In *A Room of One's Own* (1929), Woolf characterises the post-war period through the absence of a pervasive hum or vibration, a lack of resonance. In *Mrs Dalloway* we see the attempt to recalibrate the idea of echo and resonance and to listen in to and celebrate urban, mechanical noise in order to renew or reconceive its associations with injury and shock. While this is a novel that revels in the connective possibilities of urban sounds, its depiction of the medical profession chimes with those doctors who would go on to campaign against the dangers of noise and for whom the sonic legacy of the war produced a preoccupation with the injurious potential of mechanical sound. Wasted energy was at the heart of their concerns. In her depiction of Septimus's Harley Street neurologist, William Bradshaw, Woolf is at her most excoriating. Bradshaw's mantra is efficiency and proportion; anyone deemed transgressive or 'unfit' is subject to silencing or incarceration: 'Naked, defenceless, the exhausted, the friendless received the impress of Sir William's will. He swooped; he devoured. He shut people up.'[52] The character of Bradshaw is conventionally understood to be based on Sir Maurice Craig, a psychiatrist and nerve specialist who treated Woolf for more than two decades. Craig was on the 'War Office Committee of Enquiry into "Shell Shock"' and was also one of the doctors engaged in interwar noise abatement work. In particular, he was part of a 1928 British Medical Association delegation to Parliament to have noise classified as a nuisance under a Public Health Act.[53] Alongside him was Thomas Horder, who would go on to found the Anti-Noise League in 1933. Neville Chamberlain (then Health Minister) was not convinced: he thought the medics were exaggerating and understood response to noise as a function of individual mental health. Although the delegation did not succeed in its aims, this moment crystallises a wider interwar movement to legislate against noise.

Horder was a draper's son, who was taught by H. G. Wells at a crammer in Red Lion Square before training at Barts Hospital. He went on to become physician to various royals (after his famous diagnosis of Edward VII in 1910), prime ministers and London high society from his offices in Harley Street. Horder and his colleagues at the NAL campaigned endlessly to raise awareness of the dangers of noise, being 'belligerent' about noise, as he put it.[54] Noise complaints rose through the 1930s, and the NAL laid claim to a range of noise legislation such as the nighttime banning of motor horns and other zoning practices. At the organisation's Noise Abatement Exhibition at the Science Museum in 1935, 44,000 visitors were educated in what was called the 'brutality of unnecessary noise' and encouraged to 'get civilised; keep silent'.[55] In his opening address at the Exhibition, Horder praised a newly 'noise-conscious' nation:

> there is no more need to put up with this stupid expenditure of nerve energy, through needless noise, than there is to put up with the expenditure caused by bad housing, tainted food, impure water or any other of the preventable conditions that destroy health.[56]

While H. G. Wells addressed the crowd by praising noise, noting that he would opt for the current soundscape rather than a silent world, he felt the aim was 'not to banish noise, but to control it.'[57]

The Exhibition showcased new inventions, such as noise reduction technologies for the home and workplace including silent typewriters, quiet flush toilets and vacuum cleaners that could be operated around the beds of sleeping hospital patients. As James Mansell has argued, the NAL sought the input of modern professionals such as architects, engineers and designers to reconfigure modern technology to reduce its sonic byproducts.[58] The focus was overwhelmingly on mechanical noise and human control of technology. In Horder's words: is 'man to be mutilated by the machines which he has invented'?[59] He conceded that scientific evidence about the physical effects of urban noise on the body remained inconclusive and instead employed a language of 'nerve force' and the entropic decay of energy in the human and national machine. The ear was always the locus of danger. The NAL argued that bodies and minds were being bombarded and depleted by invisible waves in the ether, and they saw it as their job to educate and inform. According to Horder, noise, while not inducing organic disease apart from deafness, opened the door to disease by depressing vitality and wearing 'down the nervous system [. . .] the master stuff of our bodies, and it does stultify our spirit.'[60]

The solution proposed repeatedly by the NAL was public education in sonic self-regulation. The narrative of catastrophe and decline so prevalent in the 1930s could be reversed through teaching people to be quieter (with the repeated refrain of 'needless' or 'unnecessary' noise). The NAL's doctors focused on the impact of noise on heart rate, metabolism and sleeping patterns, but the language quickly shifted to an interwar language of efficiency and behaviour: a language articulated in the context of the medical profession in *Mrs Dalloway*. The focus on efficiency found in the work of F. W. Taylor and his theories of scientific management and labour productivity was utilised by the noise abaters to advocate for the individual and national benefits of a quieter society. This was a language that became more insistent as the 1930s progressed. Speaking in 1939, Horder linked noise abatement work to war preparedness:

> as never before – does this nation need cool judgement, steady nerves and reserves of force. The untidiness of our civic lives and our health-arrangements are sapping these vital things and, among the chief causes, we vaguely say 'greater nervous strain of modern life' is responsible. Much of this is due to noise outside and inside our homes.[61]

Aside from the 1939 Control of Noise Order that regulated the wartime soundscape, Horder's rhetoric made keeping quiet a civic duty. The sonic dimension of the interwar preoccupation with efficiency and waste is often overlooked.

Despite the campaigners' gestures towards the ubiquity of the noise problem and its public health rhetoric, the real targets of the campaigns are not hard to identify. One contributor to *Quiet* writes that 'efficient, hard working, successful people are not noisy or boisterous'; another that 'noise is useless: it does no good. It is a parasite, and all parasites ought to be wiped out.'[62] Horder's presidency of the Eugenics Society coincided with his chairmanship of the NAL, and the classist biopolitics of his speeches targets a working and rising middle class comprised of 'primitive and immature citizens'.[63] Horder's emphasis on 'acoustic-civilisation' and the designation of noise as a sign of 'lingering savagery' connects to a long history of associating noise-production with the working class.[64] Noise, then, becomes not just a byproduct of modernity but a mode of denigrating and regulating particular behaviours and leisure activities. On one level, the NAL's rhetoric chastised noise makers perceived as in breach of characteristics that supposedly defined Englishness. On another, their discourses posited noise as both the cause and sign of mental ill health. Horder wrote: 'our aim is equanimity: *we want to explore the hygienic value of a quiet mind* [. . .] equanimity, control, security.'[65] This resonates with Leonard Woolf's account of Virginia's doctor's advice: 'Equanimity – equanimity – practise equanimity, Mrs Woolf.'[66] In *Mrs Dalloway*, we find the same discourses of proportion, stability and control used in relation to Septimus and Clarissa: he is ultimately prescribed the rest cure, and she is relegated to her upstairs room and 'an hour's complete rest after luncheon.'[67] Woolf presents a society which views both characters as aberrant and non-productive.

Reading the novel in the context of noise abatement discourse reveals the ways in which languages of efficiency bleed across distinct areas of interwar biopolitics. As Michel Foucault outlines in *Society Must Be Defended*, the rise of biopower indicates a shift away from individualised disciplinary techniques of control and surveillance towards a collective focus on political control at the level of population or man-as-species. In *Mrs Dalloway* we see both processes at work: a combination of individualised disciplinary technologies that manipulate 'the body as a source of forces that have to be rendered both useful and docile', and the biopower associated with a political regulation of life that aims to 'achieve an overall equilibrium that protects the security of the whole from internal dangers.'[68] Septimus and Clarissa are both subjected to enforced rest and confinement (the threat of the rest cure leads Septimus to his death), but Bradshaw's wider project of conversion and proportion is a nationalist and imperialist eugenic programme of control. While noise abatement work was not about the curtailment of life, it *was* about the notion of national homeostasis that Foucault discusses in relation to population control. It employed a medicalised language of public hygiene and the protection of the human from the deleterious effects of the urban environment: one of the starting points of biopolitics, as Foucault indicates.[69] Noise purportedly caused the depletion and deterioration of national energy reserves; therefore, a sonic biopolitics aimed to educate the population on the noise threat through a combination of state intervention and training in self-regulation. A language of 'acoustic nationalism', or the creation of a silent, docile nation through sonic hygiene, emerges in this period in relation to other ways of narrating Britishness. It originates out of a post-war context of concern around extreme, high-decibel noise but quickly becomes a provocative touchstone for other anxieties.

Woolf was to continue her fascination with the interplay between sound and other regulatory discourses. *The Years* (1937), for example, offers another raucous urban soundscape but also interrogates the place of popular and art music in defining communities and nations; Beryl Pong takes up this story in Chapter 5 of the current volume.

To conclude the present chapter, I want to stop for a moment on the attention to *rural* noise in Woolf's final novel *Between the Acts* (1941) and to suggest an ongoing connection to Horder and noise abatement work. The NAL were affiliated with and worked alongside the Council for the Preservation of Rural England; together, they constructed a quiet rural environment in the service of national identity. In his edited volume *Britain and the Beast* (1937), preservationist Clough Williams-Ellis declared a desire to 'reverse the effects of a nation where 'disorder, ugliness and inefficiency are generally accepted and tolerated.'[70] One of the volume's contributors was Horder, for whom quiet and the revolt against din were central to the preservation of the 'Amenities Front'. The spaces of rural refuge were invaded not only by new housing and advertising but by the supposed rowdy excesses of urban day trippers. Horder made the particulars of his scapegoat very clear: 'the sports car with the flamboyant exhaust . . . generally driven, I think, by a young gentleman in suede shoes with marcelled hair.'[71] Noise was perceived to be one of the disfiguring and polluting contaminants of the rural. And this is made more apparent by the sonic dimension of the preservationist debates; rural noise was seen as a pollutant akin to litter and advertising. The rural was both a place to escape noise and a symbol of its all-pervasiveness. An idealised, silent rural was the backdrop of the NAL's campaigns; the covers of their magazine, *Quiet*, featured idealised and depopulated country scenes. The rural occupied a complex position in narratives about interwar noise, both as fetish, space of healing and retreat but also as a space subject to the same regulatory discourses as the urban. The countryside held the potential for sonic healing if the right sounds were listened to in the right way.

Critics have recognised *Between the Acts*' engagement with rural preservation debates in the novel-play's references to rural amenities and interwar house building.[72] The novel-play's sonic heterogeneity, however, also engages with these battles over the definition of the 'countryside'. *Between the Acts* could be considered noise art in its constantly shifting attention to production and reception of disruptive sounds. The text registers a plethora of human, natural and mechanical sounds: from the 'noise of china and chatter', the 'drone of the trees', the blaring and ticking of the gramophone and the 'bellow' of cows 'sounding loud in the ear of the present moment.'[73] This culminates in the 'Present Day' section of the village pageant, a high-decibel apex rendered in jagged, fragmented prose poetry: 'What a jangle and a jingle!'[74]

At the broadest level, again, Woolf seems interested in a cacophonous countryside complete with the portable gramophones railed against by the preservationists. She embraces sonic excess by depicting not a retreating or isolated selfhood or nationhood but an acoustic ecology. The novel's 'gigantic ear' is on the listen-out for the ways in which a shift in earspective can reveal the sounds of the natural world. The countryside becomes not a static object of aesthetic contemplation but a reverberating, dynamic ecology. As Miss LaTrobe, the pageant's director, sits contemplating her next creation at the close of the novel, we learn that she 'listened': with sight obscured 'there was no longer a view – no Folly, no spire of Bolney Minster. It was land merely, no land in particular.'[75] Words rise to the surface as sound.

Noise, however you parse it, connects to sonic materiality. Attending to the acoustics of literary modernism is not just about registering new soundscapes and ways of hearing and making sound but also about the way sound connects and regulates bodies. Literary fiction played a central role in registering and shaping narratives about noise. The soundscape of *Mrs Dalloway*, in its non-discriminating attention to all sounds, its registering of the sonic legacies of World War One, and its interest in sonic modes of connection, is part of much wider interwar debates about the impacts of noise, particularly the biopolitics of acoustic regulation and demands for a quiet citizenry.

Notes

1 R. Murray Schafer, *The Soundscape: Our Sonic Environment and the Tuning of the World* (Rochester: Destiny Books, 1994), p. 4.

2 See, for example, https://citiesandmemory.com/covid19-sounds/, and The Silent Cities Project (https://osf.io/h285u/).

3 A. H. Davis, *Noise* (London: Watts & Co., 1937), p. 5.

4 Dan McKenzie, *City of Din: A Tirade against Noise* (London: Adlard & Sons, 1916), p. 34.

5 Aldous Huxley, 'Silence', in *The Perennial Philosophy* (London: Chatto & Windus, 1947 [1945]), pp. 247–50 (249).
6 Douglas Kahn, *Noise, Water, Meat: A History of Sound in the Arts* (Cambridge: MIT Press, 1999), p. 9.
7 Angela Frattarola, *Modernist Soundscapes: Auditory Technology and the Novel* (Gainesville: University of Florida Press, 2018), esp. chapter 4.
8 Virginia Woolf, 'A Sketch of the Past', in *Moments of Being*, ed. Jeanne Schulkind, 2nd edn. (New York: Harcourt Brace, 1985), pp. 64–159 (67).
9 Virginia Woolf, 'Character in Fiction', in *The Essays of Virginia Woolf*, 6 vols, ed. Andrew McNeillie (London: The Hogarth Press, 1988), III, pp. 420–38 (433–4).
10 Ibid., p. 435.
11 Virginia Woolf, 'Thoughts on Peace in an Air Raid', in *The Essays of Virginia Woolf*, 6 vols, ed. Stuart N. Clarke (London: The Hogarth Press, 2011), VI, pp. 242–8 (243).
12 See, for example, Melba Cuddy-Keane, 'Virginia Woolf, Sound Technologies, and the New Aurality', in *Virginia Woolf in the Age of Mechanical Reproduction*, ed. Pamela L. Caughie (London: Garland, 2000), pp. 69–96; Pamela Caughie, 'Virginia Woolf: Radio, Gramophone, Broadcasting', in *The Edinburgh Companion to Virginia Woolf and the Arts*, ed. Maggie Humm (Edinburgh: Edinburgh University Press, 2010), pp. 332–47; Kate Flint, 'Sounds of the City: Virginia Woolf and Modern Noise', in *Literature, Science, Psychoanalysis, 1830–1970: Essays in Honour of Gillian Beer*, ed. Helen Small and Trudi Tate (Oxford: Oxford University Press, 2003), pp. 181–94; and Frattarola, *Modernist Soundscapes*, chapter 3.
13 Douglas Kahn, *Noise, Water, Meat*, pp. 2–3. See also Sam Halliday, *Sonic Modernity: Representing Sound in Literature, Culture and the Arts* (Edinburgh: Edinburgh University Press, 2003) and *Sounding Modernism: Rhythm and Sonic Mediation in Modern Literature and Film*, ed. Helen Groth, Julian Murphet and Penelope Hone (Edinburgh: Edinburgh University Press, 2018).
14 See, for example: James G. Mansell, *The Age of Noise in Britain: Hearing Modernity* (Urbana: University of Illinois Press, 2016); Emily Thompson, *The Soundscape of Modernity: Architectural Acoustics and the Culture of Listening in America, 1900–1930* (Cambridge: MIT Press, 2002); and Karin Bijsterveld, *Mechanical Sound: Technology, Cultural and Public Problems of Noise in the Twentieth Century* (Cambridge: MIT Press, 2008).
15 James G. Mansell, 'Noise', in *Sound and Literature*, ed. Anna Snaith (Cambridge: Cambridge University Press, 2020), pp. 154–69 (155).
16 Virginia Woolf, *The Diary of Virginia Woolf*, ed. Anne Olivier Bell, 5 vols (London: The Hogarth Press, 1980), II, p. 299.
17 Woolf, 'A Sketch of the Past', p. 72.
18 McKenzie, *City of Din*, p. 25.
19 Norman McLachlan, *Noise: A Comprehensive Survey from Every Point of View* (Oxford: Oxford University Press, 1935), p. iii.
20 Allard Dembe, *Occupation and Disease: How Social Factors Affect the Conception of Work-Related Disorders* (New Haven: Yale University Press, 1996), p. 168.

21 See the reports of the IHRB: 'Effect of Noiseless Typewriters', *Industrial Health Research Board Report*, 19 (1938); and Millais Culpin and May Smith, 'The Nervous Temperament', *Industrial Health Research Board Report*, 61 (1930).
22 Virginia Woolf, *Mrs Dalloway*, ed. Claire Tomalin (Oxford: Oxford University Press, 1992), p. 5.
23 Ibid., p. 3.
24 Ibid.
25 Ibid., pp. 48–9.
26 Virginia Woolf, 'Heard on the Downs: The Genesis of Myth', in *The Essays of Virginia Woolf*, 6 vols, ed. Andrew McNeillie (London: The Hogarth Press, 1987), II, pp. 40–2 (40).
27 Charles Myers, *Shell Shock in France 1914–1918* (Cambridge: Cambridge University Press, 1940), p. 13.
28 Edwin Ash, *The Problem of Nervous Breakdown* (London: Mills and Boon, 1919), pp. 269–70.
29 Woolf, *Mrs Dalloway*, pp. 17, 16.
30 Virginia Woolf, *The Diary of Virginia Woolf*, ed. Anne Olivier Bell, 5 vols (London: The Hogarth Press, 1977), I, p. 32.
31 Paul K. Saint-Amour, *Tense Future: Modernism, Total War, Encyclopedic Form* (Oxford: Oxford University Press, 2015), pp. 7–8.
32 Woolf, *Mrs Dalloway*, p. 25.
33 Ibid., p. 197.
34 Ibid., p. 62.
35 Elizabeth Bowen, *To the North* (Harmondsworth: Penguin, 1984), p. 41. Thank you to Imogen Free for this reference.
36 Yaron Jean, 'The Sonic Mindedness of the Great War: Viewing History through Auditory Lenses', in *Germany in the Loud Twentieth Century*, ed. Florence Feiereisen and Alexandra Merley Hill (Oxford: Oxford University Press, 2012), pp. 51–62 (53).
37 Ford Madox Ford, 'Letter to Joseph Conrad: September 1916', in *The Presence of Ford Madox Ford: A Memorial Volume of Essays, Poems and Memoirs*, ed. Susan Strang (Philadelphia: University of Pennsylvania Press, 1981), p. 172. See also Tom Vandevelde, 'Are You Going to Mind the Noise?: Mapping the Soundscape of *Parade's End*', in *Ford Madox Ford and 'Parade's End': The First World War, Culture, and Modernity*, ed. Ashley Chantler and Rob Hawkes [*International Ford Madox Ford Studies*, 13] (Amsterdam: Rodopi, 2014), pp. 53–66.
38 Ford, 'Letter to Joseph Conrad', p. 173.
39 Ibid., pp. 172–3.
40 Ibid., p. 173.
41 Woolf, *Mrs Dalloway*, p. 117.
42 Ibid., p. 188.
43 Ibid., pp. 31 and 182.
44 Ibid., p. 183.
45 Ibid., p. 190.
46 Ibid., p. 89.
47 Ibid., p. 28.

48 Ibid., p. 101.
49 See John Picker, *Victorian Soundscapes* (Oxford: Oxford University Press, 2003) and Veit Erlmann, *Reason and Resonance: A History of Modern Aurality* (New York: Zone Books, 2010).
50 Woolf, *Mrs Dalloway*, p. 28.
51 Ibid., pp. 243 and 241.
52 Ibid., p. 133.
53 See Mansell, *The Age of Noise*, pp. 37–8.
54 Thomas Horder, 'Forward', in *Noise Abatement Exhibition: Science Museum, South Kensington, 31st May–30th June 1935* (London: Anti-Noise League, 1935), p. iii.
55 Ibid.
56 Horder, 'The Elimination of Needless Noise', in *Noise Abatement Exhibition*, pp. 1–3 (3).
57 For more on Wells's contribution to the event, see Mansell, *The Age of Noise*, pp. 54–6.
58 Ibid., pp. 58–9.
59 Horder, 'The Elimination of Needless Noise', p. 1.
60 Thomas Horder, Horder Archive, Wellcome Collection, GP/31/B.4/4, typed speech. I am grateful to the Wellcome Collection for permission to quote from the Horder Archive.
61 Horder, Horder Archive, Wellcome Collection, GP/31/B.2/24, 9.
62 *Quiet* (July 1937), 23.
63 Horder, 'The Elimination of Needless Noise', p. 3.
64 Horder, Horder Archive, Wellcome Collection, GP/31/B.2/24, 3; and *Quiet* (June 1936).
65 Horder, *Silencing a Noisy World*, Anti-Noise League – General File, 1934–38, National Archives HLG/52/1210.
66 Leonard Woolf, *Downhill All the Way: An Autobiography of the Years 1919–1939* (New York: Harvest, 1989 [1967]), p. 51.
67 Woolf, *Mrs Dalloway*, p. 156.
68 Michel Foucault, *Society Must Be Defended: Lectures at the Collège de France, 1975–6*, ed. Mauro Bertani and Alessandro Fontana, trans. David Macey (New York: Picador, 1997), p. 249.
69 Ibid., p. 245.
70 Clough Williams-Ellis, *Britain and the Beast* (London: J. M. Dent, 1937), p. xv.
71 Horder, Horder Archive, Wellcome Collection, GP/31/B.2/24.
72 See Mark Hussey, *'I'd Make it Penal': The Rural Preservation Movement in Virginia Woolf's Between the Acts* [Bloomsbury Heritage] (London: Cecil Woolf, 2011), 15pp.; and Clara Jones, 'Virginia Woolf and "The Villa Jones" (1931)', *Woolf Studies Annual*, 22 (2016), 75–96.
73 Virginia Woolf, *Between the Acts*, ed. Gillian Beer (London: Penguin, 1992), pp. 64, 11 and 85.
74 Ibid., p. 109.
75 Ibid., pp. 125 and 126.

Earpiece 1
'Feel Dumb. Don't Cry'
Inside a Soundproof Grey Room

Jaipreet Virdi

> *"You will say . . .*
> *Bad day –*
> *soundproof*
> *gray room.*
>
> *No hear*
> *"cowboy,*
> *baseball,*
> *hot dog."*
>
> *Feel dumb.*
> *Don't cry.*
> *Be numb."*[1]

I open the door and enter the office, almost stumbling my footing on the discarded toys on the carpeted floor. The receptionist greets me, takes my name, and instructs me to take a seat. I look around, taking in how little has changed in all the years I've been coming here: manufacturer posters on the wall showcasing a smiling child wearing their hearing aid; displays of colorful and patterned options for hearing aid models; framed clippings of the office's annual 'Hear Here' summer festivals. The four rooms, an office for the audiologist, the fitting room, and the offices of Chris and Nick, my hearing aid specialists who have known me for over twenty-five years.

And there on the wall, overlooking all of us, a portrait of Peter Keller, who founded the office in the 1950s in downtown Toronto. Established in 1922 as Acousticon of Toronto, it was the first to provide hearing aid services, expanding to pediatric audiometry and instrument fitting when Keller took over.[2] The office additionally offered services to local primary schools – including mine – and my childhood was marked by the elderly Mr. Keller's (as I've always known him) regular visits for ear exams and

DOI: 10.4324/9780429291616-5

earmold fittings. Chris and Nick joined the company in anticipation of Keller's retirement, but their school visits stopped once government cutbacks eliminated the program. From then on, to obtain their services, I visited their office in the city.

The visits are predictable. I'm examined by either Chris or Nick, who begin by peering into my ear with the otoscope, followed by a mechanical inspection of my hearing aids, and, if needed, cleaning and parts replacement. We chat lively, as I tell them about my studies and interests. Their smiles radiate as they insert the silicone into my ear canal to shape new molds. I sit in silence for some time as the silicone dries, then they remove it, put it in a labelled box, and give me instructions for picking it up once it's ready. I leave the office and continue with my day.

Occasionally, I'm told I need an updated audiological assessment. This is a test of how well I hear – or, more accurately, how much I *don't* hear – measured by an audiometer and documented on an audiogram: a graph visualizing the test results across a range of frequencies (pitches) and sound intensity (loudness) that a person is able to hear.[3] The audiogram provides the audiologists with an objective insight into my deafness, one that they can then use to make an assessment for which hearing aid models best fit my specific needs, especially to target my speech frequency range. Though I've come to understand the audiogram as a prescription of sorts, the feeling of failure that comes from seeing the x's and o's across the range of what's classified as 'profound deafness' is hard to forget.

The test takes place in a grey soundproof room – or, sometimes, merely a booth – that interweaves past and present, acting as an echo chamber for all the tones emitted within its four walls. Walking through the double-walled steel doors, I encounter a familiar scene: a chair, sometimes two; a glass window; a pair of headphones hanging on the wall; speakers posted on two corners high on the ceiling; sometimes a plush animal or stickers on the wall. The window in front of me showcases the audiologist seated at the table, surrounded with machines, binders, and papers. She is maybe a meter on the other side of the window but seems to be further away – in another dimension, perhaps, on the outside while I am confined in the grey soundproof room.

It is a room that makes one feel insignificant. A lonely space with an absence of sound. I fix my vision on the window, longing for a connection to the only other human, clinging on to this need. I am afraid to let go, lest I be trapped in the hushed abyss of the room and enveloped into nothingness.

As instructed, I take off my hearing aids and put on the headphones. After some time, sound reverberates through my skull, adjusting in intensity and frequency until I can clearly make out the audiologist's voice. 'Do you hear me?' she asks. I nod, then clear my throat and say 'Yes.' 'Okay, good, we're going to begin.' A series of instructions follows, the order unchanged in the thirty-plus years I've been receiving these tests. We begin, first with the sound test. A series of beeps, of various frequencies, noises that stay in the recesses of my mind. I hold on to the clicker, palm sweating. I worry and think of failure. Thoughts quickly go through my mind, and I worry and worry and worry about what I cannot hear.

Is this a beep?
Or this?
This must be a beep.
[*click*].

I look intensely at the audiologist's face for clues, inspect her body language to ascertain when she fiddles with the machines to change the tones. She's stoic and reveals nothing. I keep looking until I suspect I see glimmers of adjustments.

Is this a noise? Or this? Is this a noise? I count the seconds in between frequencies. I click when I am sure, I click when I am not sure. I pretend I heard something; I pretend my hearing is better than it is.

[*click*].
I don't think it's a beep. Oh no.
Count the seconds, the minutes. There's a pattern to this. I need to figure it out.
[*click*].

Tears well up in my eyes. The stress of the test is overwhelming me. I cannot escape the feelings of failure. My mind races and struggles to find clues, any clues, in her face and her body so I can work around the test. I need signals. I need more than noise. I need to understand. My ears don't work on sound alone.

[*click*].
Will she know that I don't know?
Don't cry. Don't cry.
[*click*]. [*click*]. [*click*].

The first part of the test eventually ends. I'm never told how well I do. All I know is that the charges of lines and x's and o's on the audiogram are a measure of my failure to hear.

I put my hearing aids back on. We then move on to the speech-recognition test, or sometimes the speech-reception threshold, an assessment of how well I hear and understand words. *Say Baseball. Say Hotdog. Say*

Cowboy. Say Door. Say Goose. Say Chalk. For years I've tried to memorize the words – again, looking for a pattern by which I can ace the test and present myself as being able to hear better than I actually can. I can see the audiologist, but her mouth is hidden behind a piece of paper. She knows if I see her lips shaping the words, her jaw and tongue movement, I will hear more. The child in me viewed this as a deception: why bother with this test when I do not experience sounds and speech this way, with obscured mouths and static faces and body language? I seek out clues that are not hidden by the cardboard. I look into her eyes, I watch for muscle twitches. I ask her to repeat.

Most of the time, despite my best attempts, I cannot recognize a word or make sense of it through the blurred noise. My head hurts from trying, and the tears are replaced by heated flames of anger and frustration. The audiologist pulls down the paper, and I can see her lips. We do the test again. It moves far more quickly. *Baseball. Hotdog. Cowboy. Pool. Chalk. Goose. Page. Shout. Hash.*

After a few minutes, the test is complete. The audiologist then opens the door to the grey soundproof room. The last time I had this test, she was astonished and tells me why: with lip-reading, I identify 98% more words than when I hear through my ears alone; through my hearing aids – just my hearing aids – I manage to understand less than 50% of what is being said. What are my ears, then, but not my eyes? My brain makes sense of noise through both aural and visual clues, filling in the blanks of what is being missed, to construct as accurate a picture as possible of what is being said. It is at this moment I come to realize that this spectrum of abilities – a range of hearing possibilities – is what makes me empowered. Despite what I experience in the grey soundproof room, I do not hear with my ears alone. Nor do I need to.

Notes

1 Brenda Jo Brueggemann, 'An Assessment of the Speech-Recognition Threshold' (extracts), in *Lend Me Your Ear: Rhetorical Constructions of Deafness* (Washington, DC: Gallaudet University Press, 1999), p. 147.

2 This discussion of Peter Keller was originally published in Jaipreet Virdi, *Hearing Happiness: Deafness Cures in History* (Chicago: University of Chicago Press, 2020), pp. 258–9.

3 On the histories of the audiometer and audiogram, see: Viktoria Tkaczyk, Mara Mills, and Alexandra Hui (ed.), *Testing Hearing: The Making of Modern Aurality* (Oxford: Oxford University Press, 2020); Coreen McGuire, *Measuring Difference, Numbering Normal: Setting the Standards for Disability in the Interwar Period* (Manchester: Manchester University Press, 2020); and Jonathan Sterne, *The Audible Past: Cultural Origins of Sound Reproduction* (Durham, NC: Duke University Press, 2003).

4 Automatic Voices

Modernism, Telephony, and Delusion

Andrew Gaedtke

Modernist fiction is replete with voices whose sources and ontological status are often ambiguous to both character and reader. Innovative narrative techniques for representing characters' lived experiences and thought processes emerged at a historical moment when voices had become technologically detached from persons. As Friedrich Kittler has argued, the gramophone, telephone, and radio made voice-hearing an experience that was not limited to the mentally ill, and these media offered useful ways to account for one's experiences of auditory hallucination.[1] Modernism's media ecology reconfigured the meaning of voice-hearing and eroded the phenomenological differences that would mark off 'normal' from 'pathological' experiences. Modern fiction often captured this ambiguous convergence of voice-hearing, technological media, and delusional self-narration in ways that may be instructive for understanding the lived experiences of voice-hearing.

Among the most striking fictional representations of this convergence is the 1941 novel *Hangover Square* by Patrick Hamilton. The novel is focalized through George Harvey Bone, a marginal member of a group of hard-drinking Londoners that is organized around an aspiring actress named Netta Longden. Bone's hopeless obsession with Netta is answered by her increasingly cruel treatment of him, and he progressively descends into a dissociative and perhaps delusional state that ends in her murder. If the title of the novel gestures toward Bone's alcoholism, his struggle with mental illness is signaled by an epigraph that provides a clinical definition of schizophrenia.[2] The original, complete title, *Hangover Square; or, The man with two minds; a story of darkest Earl's Court in the year 1939*, might signal a hybrid genre. It is a work of fiction that also resembles memoirs of mental illness and narrative case studies, such as Alexandre Luria's *The Man With a Shattered World* (1972), which attempts to capture the altered phenomenology that results from a severe neurological disorder. While invoking the phenomenological sense of 'world', the

DOI: 10.4324/9780429291616-6

subtitle also signals the historical specificity in which that psychic space is forged: *Earl's Court in the year 1939*. Within this historical moment certain material and discursive resources are available to Bone for making sense of his altered state. More specifically, his anxious and frustrated attempts to connect with a distant other are often carried out via the telephone, and the device becomes the primary channel through which his experiences of voice-hearing, depersonalization, and psychic influence are given form.

What distinguishes the formal technique of *Hangover Square* is Bone's episodic experience of unreality. The transition into this altered state is often marked by a distinct mental 'click' followed by a radical transformation of the structure of his experience. Most conspicuously, persons and objects appear as unreal simulacra of themselves. Bone often invokes audio and visual media in order to make sense of this artificiality of such experiences and the distinct mental 'click'. It is the click heard when the other picks up or puts down the phone at the other end of the line: a connection to the other, a shared world, is made or lost. The experience of the telephone becomes paradigmatic of Bone's altered states.

Bone meditates on the telephone booth as a space of both hermetic enclosure and disembodied action at a distance that resembles his own hermetic mental space that is cut off from the world of consensus.

> In the line of telephone booths there were a few other people locked and lit up in glass, like waxed fruit, or Crown jewels, or footballers in a slot machine on a pier, and he went in and became like them – a different sort of person in a different sort of world – a muffled, urgent, anxious, private, ghostly world, composed not of human beings but of voices, disembodied communications – a world not unlike, so far as he could remember, the one he entered when he had one of his 'dead' moods.[3]

The booth becomes a third space that is neither the 'normal' world of consensus nor the world experienced by Bone during his '"dead" moods' but functions as a portal between the two. Even when he is not in his altered state of unreality, he may enter the telephone booth and experience many of the phenomenological features that define that state: hermetic enclosure, the depersonalization of others, disembodied voices, and uncanny influences and communications at a distance. If the mental 'click' initially serves to punctuate the stark phenomenological differences between the normal world and the other world of Bone's delusional state, the unreal quality of communications media such as the telephone creates an intermediate world – a space of continuity between the two worlds. When Netta picks up his call, the 'click' of the receiver underscores this trans-world effect:

> There was the longed-for click – a click which relieved the tension in his heart in rather the same way as the click in his head relieved him from one of his 'dead' moods – and he heard her say 'Hullo! . . . '[4]

As David Trotter has shown, there is evidence that Bone's experience of the telephone box as an uncanny space of enclosure and intimacy was not entirely unusual. Trotter cites an article in the *Daily News* that describes the space in rather similar terms:

> When this closes upon him he is hermetically sealed, and might rave himself hoarse without conveying the faintest sound to anyone outside. There is a pane of glass in the door, so that the officer in charge could see the inmate of the box, though unable to hear him. [. . .] The telephone box is a wonderfully compact and all but perfect little institution, the only drawback being that it is entirely unventilated, and the person sitting in it is literally hermetically sealed up.[5]

The telephone box is a topos of psychic extremes that is well suited to represent the phenomenology of psychotic voice-hearing: it is a space of hermetic enclosure in which one is cut off from the shared work of social consensus, and one may 'rave himself hoarse' without reaching others. The Other at the end of the line is a compensatory interlocutor who takes an altered form: at once intimate and distant, a familiar and yet mechanical in voice.

During his frequent calls to Netta, the telephone does not simply become a channel of intimacy; its thick mediation of the voice through a crackling and failing line manifests a desiccated and mechanical presence. Trotter writes that, in modernist literature, the opacity of dialogue between characters speaking on the telephone is often 'already rendered opaque by fizzing static and circumambient noise. Modernist phones simply do not work, in the most basic technical sense; even when they do work, they don't, because their working is itself an estrangement, a disconnection'.[6] When he encounters Netta in person during one of his 'dead moods', he hears her voice as mechanical and detached as if through a noisy telephone line.

> He knew it was Netta well enough – but it was a different Netta. Although he could see her she was remote, almost impalpable, miles away – like a voice over the telephone, or the mental construction of the owner of a voice one might make while phoning – a ghost, if you liked.[7]

The Netta whom he experiences in his altered state is, like a voice on the telephone, a simulacrum. The distinction between real and unreal voices, thoughts, and persons becomes unclear in his altered state and in the telephone booth.

Bone feels that he has become little more than a device subjected to the will of Netta – an other whose name implies a 'net' or network in which he has been incorporated:

> It was loaded, overloaded with voluptuous yet subtle intimations of her personality. Netta. The tangled net of her hair – the dark net – the brunette. The net in which he was caught – netted. Nettles. [. . .] In her net. Netta. You could go on like that for ever – all the way back to London.[8]

This verbal network of associations becomes an element of his inner monologue. When he finds himself disturbed by a stranger wearing a hairnet, Bone thinks,

> Net . . . It dawned on him. Of course, that was it. Net. Which equalled Netta . . . He had been quite right: the woman *had* been hurting him – she had been trying, all the time as she talked to her friend, to remind him of Netta.[9]

Netta becomes not only a singular object of desire for Bone; the name also signifies networks of communication in which Bone has become ensnared to the point that he finds himself 'obscurely tortured' by any reminder of the word 'net' and its associations. In this way, the telephone itself becomes a torture device:

> she had made this instrument, the telephone, so terrifying and odious that he could no longer use it for any purpose without a feeling of trepidation, without a feeling that the person at the other end of the line was going to hurt him.[10]

While Netta's cruel treatment of Bone over the telephone is the ostensible source of his anxiety, the device also becomes a hypostasization of obscure and potentially persecutory communication through its transmission of voices that are at once not real enough and overwhelming.

Bone becomes convinced that Netta alters his mental states through electromagnetic transmissions that are not limited to those passed through the telephone line.

> There was, he knew now, a definite sphere of sexual attraction, a halo, a field of physical and magnetic influence, which Netta carried about with

> her wherever she walked. This invisible yet palpable influence petered out at a certain distance, about two feet away from her body, which was its centre and source. So if he kept out of range, if he was more than two feet away from her, he was secure from its effect. But if he went towards her and into it, or she came towards him and either consciously or unconsciously smote him with it, he was made to suffer indescribable things.[11]

Despite her cruel treatment of him, he continues to serve her to the point of humiliation. His description of her 'field of physical and magnetic influence' wavers between a common metaphor for physical attraction and the more literal, causal logic of delusional experience, and reflects a tendency in schizophrenic discourse to collapse metaphorical meaning into literal experience.

The attempt to capture the phenomenology of psychosis through comparisons to technological media such as telephone, radio, and film is consistent with many self-reports of schizophrenic patients and voice hearers who experience a feeling that the elements of their world are conspicuously rendered for them by a process that is often compared to that of a machine. Carl Jung's famous case study, 'Analysis of a Case of Dementia Praecox as a Paradigm', describes a patient who feels that her nervous system is comprised of a telephone exchange which communicates intermittently. Within this delusion, the patient believes that the voices she hears are those of the telephone itself. This includes a running commentary on the behavior of both doctor and patient: 'The doctor should not bother himself with these things.'[12] Elsewhere, we have a direct critique of the patient's evasive tendencies. 'Telephone: You're leading the doctor round the whole wood.'[13] The voice of 'Telephone' occupies a third position, beyond and between doctor and patient, in which the unsaid can be said without culpability in the mechanical voice that is at once internal to the patient and yet displaced onto no-one in particular. Avital Ronell argues, 'The telephone acts as a narrator, earwitness, and interpreter for the irritated couple; it establishes a dimension of thirdness which every couple, in order to get somewhere, requires.'[14] In this instance, the device's mediation between self and other is so thick that noise becomes signal. The machine itself speaks, not some other person but a more abstract Other that speaks truths persons are unable to acknowledge.

Jung's case study was hardly unique. Emil Kraepelin, the German psychiatrist who first described the nosological category of *dementia praecox* (a precursor to schizophrenia), described patients who complained of 'magical, magnetic, electrical, physical, hypnotic forms of remote control, which are sent through all sorts of machines, telephones, galvanic batteries, sympathetic relations by invisible enemies.'[15] Daniel Paul Schreber famously explained the voices that he heard as a reception of 'nerve-rays' broadcast to him by God – a formulation that condenses the increasingly common metaphor that would link neurology and technology in clinical and psychopathology through the modern era.[16]

In her memoir of schizophrenia, *Operators and Things* (1958), Barbara O'Brien explains the voices that she hears through a delusion which closely resembles the feelings expressed by Bone in *Hangover Square*. She feels that she has been reduced to a lifeless information-processing machine – a 'Thing' that can be operated on by others like a telephone exchange or an early computer system. In her memoir, an 'Operator' explains this ontological reduction of persons to things:

> Things. Yes, of course. Think of the word with a capital initial, if you like. It may help your ego a bit. All people like you are Things to us – Things whose minds can be read and whose thoughts can be initiated and whose actions can be motivated. Does that surprise you? It goes on all the time. There is some, but far less, free will than you imagine. A Thing does what some Operator wants it to do, only it remains under the impression that its thoughts originate in its own mind.[17]

In many ways, this account reflects the flattened ontological world that Bone experiences in which persons are reduced to things that are animated not by reasons for acting but by causal, technical systems. When he lapses into his 'dead moods', Bone experiences himself and others as lifeless automata operated upon by unseen technological forces: 'Imagine it – wandering about like an automaton, a dead person, another person, a person who wasn't you, for twenty-four hours at a stretch!'[18]

The strangeness of these technological delusions in fiction and memoirs is reduced when compared to the clinical explanations of psychiatrists and neurologists of the period. Voice-hearing and delusions of thought broadcasting were often regarded by clinicians as the senseless, epiphenomenal effects of abnormal neurological systems that were often compared to dysfunctional electrical devices. Eugen Bleuler, the Swiss psychiatrist who introduced the nosological category of 'schizophrenia', wrote that 'the mutual influence of different psychic functions works like an electric device where functions are switched on and off.'[19] Such analogies were reinforced by the voices of engineers who developed telephony and radio, many of whom believed that the brain emitted detectable electromagnetic impulses and that their devices would enable the direct transmission of thoughts. Verena Kuni observes that

> Bleuler's *Elementarpsychologie* ('elementary psychology') provided not only the basis for an 'electro-technically worded psycho-pathology' that his Italian colleague Cerletti would soon use to develop electro-shock

> therapy. The experiments of Ferdinando Cazzamalli, who tried to prove interference by 'brain waves' and 'radio waves', and to specify different kinds of 'cerebral radio-waves' for patients with 'diseases of the "epileptic", "paranoid", "hallucinatory" or "hysterical spectrum"', also date from the same period.[20]

Cazzamalli was hardly alone in these efforts to technologically tune into the inner voices of other minds.

In 1893, Alexander Graham Bell, the inventor of the telephone, envisioned a form of electronic telepathy that would capture and communicate the electrical fields produced by the brains at a distance. 'Imagine two persons, one thousand or ten thousand miles apart, placed in communication electrically, in such a way that, without any spoken word, without sounding-board, key, or any bodily movement, the one receives instantly the thoughts of the other, and instantly sends back his own thoughts.'[21] Media historian John Durham Peters writes,

> [E]very major physicist involved in the development of radio flirted with the notion of wireless 'thought-transference'. In Britain, Lord Rayleigh, the leading acoustician of the late 19th century, J. J. Thomson, the discoverer of the electron, William Crookes, the inventor of the cathode ray tube, and Oliver Lodge, the inventor of the radio coherer, all took an active interest in what was called 'psychical research'. [. . .] William Barrett, an Irish physicist, mathematician, and founder of the British Society for Psychical Research, was one of the first to see the brain as a potentially wireless electrical transmitter. 'The brain', he wrote in 1882, 'might be regarded as the seat of radiant energy like a glowing or sounding body.' Oliver Lodge, who is better known in radio history, ventured the idea that brain was to mind as wire is to electricity, that is, a conducting medium for something less tangible. From the idea that the brain produces an electromagnetic halo it followed that these echoes might be readable. The notion of the psychical broadcasting of thoughts is coeval with the physical broadcasting of electromagnetic signals.[22]

The early science of telecommunications and psychiatry converged to render a vision of the brain as a technological device that could be 'tapped' like a telephone line. In this context, a patient could be forgiven for believing that her doctors were listening in on her thoughts or transmitting the voices that only she seemed to hear. It is difficult to disentangle the technological delusions of psychotic patients from the technological 'explanations' offered by their psychiatrists and neurologists.

Experiences of auditory hallucination do not occur within a historical or cultural vacuum but are shaped by the explanatory resources available to the subject. Of delusional formations, Darian Leader writes, 'there is a search for an alternative, something like a code or a formula or even *a gadget* that would bring order and meaning to the world.'[23] Often, this explanatory recourse to a 'gadget' manifests quite literally. The recurrence of certain technological causes at the centre of many modern delusions suggests that such devices provide solutions through which uncanny and otherwise unspeakable experiences can be causally explained. Such technological motifs recur with such consistency that they may be regarded as a genre of delusion.[24]

As I have argued elsewhere, the analogy of the brain to a telecommunications device manifested in many works of modern literature. In 1926, Wyndham Lewis wrote,

> [t]he contemporary European or American is a part of a broadcasting set, a necessary part of its machinery [. . .] at the pressing of a button, all these hallucinated automata with their technician-trained minds and bodies, can be released against each other.[25]

In Evelyn Waugh's semi-autobiographical novel, *The Ordeals of Gilbert Pinfold* (1957), a novelist suspects that his thoughts are monitored and transmitted to him by strange boxes that may have been devised by the BBC. Samuel Beckett's radio drama *Embers* (1959) replicates for the audience a character's experiences of auditory hallucination and voice-hearing.

In Ford Madox Ford's tetralogy *Parade's End* (1924–28), the shell-shocked Christopher Tietjens comes to feel that the components of his mind operate independently and communicate with one another through connections that are intermittent and uncontrollable. The analogy to low-fidelity media such as early telephone technology becomes explicit when Christopher recalls a statement he had made a decade earlier: 'His voice – his own voice – came to him as if from the other end of a long-distance telephone. A damn long-distance one! Ten years [. . .]'[26] Tietjens reflects on his mental faculties in language that might describe a poor connection: '"Intermittent!" was the word. "Intermittently functioning!"'[27] This phrase, '"intermittently functioning"', recurs at several points describing mind and machine alike and becomes a floating signifier for various forms of mechanical breakdown and pathological behaviour.[28]

The minds of characters in Ford's novels are represented as networks of components that function automatically and independently in ways that are often explicitly compared to communicating devices. This is especially clear during another of Valentine Wannop's unfortunate telephone calls:

> There was about the voice no human quality; it was as if from an immense darkness the immense machine had spoken words that dealt blows. She answered; and it was as if a substratum of her mind of which

> she knew nothing must have been prepared for that very speech; so that it was not her own 'she' that answered levelly and coolly [. . .][29]

The voice carried over the telephone is mechanically inhuman for its cruelty but also for its tinny, automatic quality; Valentine experiences a certain dissociation from her own response, which part of her mind epiphenomenally observes as automatic and independent of any control that she believed to have over herself. Throughout *Parade's End*, characters rarely direct their own thoughts but feel themselves to be epiphenomenal observers of mental processes over which they have little control. They are passively subject to communications that they cannot resist, as when Valentine thinks, 'One ought not to answer all telephone calls, but one does so mechanically.'[30]

In this way, *Parade's End* could be read as a work of 'psychotechnography' – a term that Steven Connor has proposed to describe the phantasmatic entanglements of mind and information media.[31] The novels go beyond questions of trauma to represent the ways that the mind and its disorders may be part of a broader ecology of compulsory communication and its failures. The particular forms of 'intermittent functioning' that are linked to a variety of brain disorders include not only deficits but excesses of communication, as in the unfortunate case of Colonel McKechnie, whose bursts of uncontrollable, hysterical monologues become additional layers of noise that Tietjens must ignore. Determining precisely what is noise and what might be communicative signal in the auditory environment of the war becomes one of Tietjens's primary challenges.

As Connor observes, the notion that thought constitutes an inner voice which may be technologically tapped manifests quite explicitly in Marie Corelli's 1911 novel, *The Life Everlasting: A Romance of Reality*:

> Thought is the voice of the Soul. Just as the human voice is transmitted through distance on the telephone wires, so is the Soul's voice carried through the radiant fibres connected with the nerves to the brain. The brain receives it, but cannot keep it – for it again is transmitted by its own electric power to other brains, – and you can no more keep a thought to yourself than you can hold a monopoly in the sunshine.[32]

When the brain is regarded as a machine which transmits thought the way that the telephone or radio transmits voices, it is often feared that the transmission is not secure and can be detected and received by others without one's control. This loss of control, whether through 'thought broadcasting' or through thought insertion, is one of the recurring features of psychotic delusions in the modern era.

If *Hangover Square* shares with these texts the project of representing modern experiences of techno-psychic influence and voice-hearing, it also dramatizes the ways that a 'man with two minds' navigates multiple worlds. In the 'other world' in which Bone finds himself spoken to by a mechanical voice, events are experienced as lacking conventional order, pattern, and predictability, and the reasons for other persons' behaviour become mysterious. This transformation of fundamental structures of experience recurs in accounts of schizophrenic delusions. Louis Sass writes,

> [i]ndeed, one of the reasons schizophrenic experience can be so difficult to describe or imagine is that it may involve fundamental transformations in the very structures of space, time, and identity, or in the nature and reality of human consciousness itself.[33]

The difficulty of achieving a clear understanding of these changes is rooted in not merely the content of delusions but in the ways that they often involve radically different structures of experience that cannot easily be communicated through conventional modes of discourse. Psychiatrist Karl Jaspers ultimately regarded some of these altered forms of psychotic experience as fundamentally 'un-understandable.'[34] Patients often develop innovative techniques for rendering the feelings of unreality, unworlding, and depersonalization that are common symptoms of schizophrenia, and a modification of narrative form in *Hangover Square* distinguishes the altered phenomenology of Bone's 'other world'. During his episodic experiences of unreality, the fairly conventional, if rather aimless, realist narration that comprises much of the novel is interrupted by detached, floridly metaphorical descriptions that approach the fractured and alienating qualities of a Surrealist text. The effect renders an altered phenomenology in which events occur with minimal narrative integration. When such an episode abruptly ends, a more conventional narrative form resumes, and Bone finds himself once again relating more easily to his environment, but he is unable to account for the time that was lost and his behaviour within that altered state. 'It was wonderful how he did things when he didn't know what he was doing. (Or did he, at the time, in some way know what he was doing? Presumably he did.)'[35] Even his thoughts during his altered state are inaccessible to him when he returns to relative normalcy, as if they had been thought by someone else or delivered via a machine from some uncanny other.

> Well, then, what *had* he been thinking about – what went on in his head when the shutter was down? What? What?
>
> It was no good. He had no idea. Not the vaguest idea. This was awful. He must try and think. He really must try and think. But what was the use of thinking? He never could remember, so why should he remember now?[36]

This inability to recognize one's own thoughts extends to experiences of others as well. A feeling of artificiality often reduces the distinction between things and persons, who are no longer recognized as entirely human or fully real but instead appear as automata that operate in an uncanny, unmotivated manner:

> He looked at passing objects and people, but they had no colour, vivacity, meaning – he was mentally deaf to them. They moved like automatons, without motive, without volition of their own. He could hear what they said, he could understand their words, he could answer them, even; but he did this automatically without having to think of what they had said or what he was saying in return. Therefore, though they spoke it was as though they had not spoken, as though they had moved their lips but remained silent. They had no valid existence; they were not creatures experiencing pleasure or pain. There was, in fact, no sensation, no pleasure or pain at all in this world: there was only himself – his dreary, numbed, dead self.[37]

Bone's refrain that the elements of this world 'had no valid existence' indicates that a phenomenological shift characteristic of schizophrenia and other forms of mental illness has occurred. Persons and objects are no longer ontologically distinguished. The routine exchange of conversation is emptied into a hollow protocol that is performed automatically by both Bone and the others. Importantly, his failed relation to others is coded in auditory language – 'he was mentally deaf to them' – suggesting that the voices that he hears are at once excessive in their hallucinatory status and deficient in their mechanical hollowness. His failure to discern the motivations or mental states of others is converted into an ontological claim that others have no mental states such as motives, beliefs, desires, or even sensations: 'they were not creatures experiencing pleasure or pain.' Elsewhere he describes the ways that language is emptied of semantic content, and this emptiness is attributed in part to a loss of context or of a conventional, phenomenological world: 'He could hear and understand the words, but for the moment he couldn't gather what they meant. They seemed divorced from any context: or at any rate he didn't know what the context was. So he didn't answer.'[38] In the absence of understanding the thoughts and intentions of others, the world is heard differently.

This failure to understand intentional states or reasons for action signals a departure from conventional folk psychology. Such experiences often lead schizophrenic patients to describe other persons as lifeless automata. Daniel Hutto has argued that 'our everyday understanding of intentional action is itself an essentially narrative practice – i.e. that "folk psychological" understanding always takes the form of constructing narratives.'[39] It is precisely this narrative construal of persons in terms of reasons for acting that is often altered or absent in delusional states. Elsewhere, Bone compares his altered states to the status of film in order to describe the unreal quality of other persons that he experiences in the absence of folk psychological predicates:

> A film. Yes, it was like the other sort of film, too – a 'talkie.' It was as though he had been watching a talking film, and all at once the soundtrack had failed. The figures on the screen continued to move, to behave more or less logically; but they were figures in *a new, silent, indescribably eerie world.*[40]

The technological motifs that recur with such consistency during Bone's 'dead moods' perform several functions. They indicate a phenomenological state that is marked by a feeling of artificiality. In addition, the mechanistic descriptions indicate an absence of folk psychological understanding. The absence of 'motive or conscious volition' in others reduces them to moving, talking things that Bone interacts with in the form of the wireless, telephone, and film. Bone's tendencies toward mechanistic causal explanations of the behaviour of others and himself manifests in place of any understanding of reasons for acting. In defining the epistemic framework that narrative world-making affords, Bruner distinguishes between 'reasons' and 'causes':

> The loose link between intentional states and subsequent action is the reason why narrative accounts cannot provide causal explanations. What they supply instead is the basis for interpreting why a character acted as he or she did. Interpretation is concerned with 'reasons' for things happening, rather than strictly with their 'causes' [. . .][41]

The 'new, silent, indescribably eerie world' which Bone experiences during his 'dead moods' is distinguished by an altered form of narration that offers causal explanations for behaviour rather than interpreting such behaviour in terms of reasons for acting. In this altered state where reasons, intentions, and meanings are elusive, a need for coherence often manifests in the form of a delusional system by which phenomena can be explained. This need to forge an order is evident when Bone constructs a web or network of string at the scene of Netta's murder. 'It was all threaded together. All the threads were gathered up. The net was complete. The net, Netta. Netta – the net.'[42] As we have seen, the net is a figure for Netta, but this relation is reversible: Netta stands in for the network of communication from which Bone often feels excluded and by which he is tormented. The net of string – like a delusion – is Bone's attempt to reconstitute a sense of order, or a world, through which things are once again connected within a system: 'the net – all complete and fitting at last.'[43]

Eugen Bleuler described the tendency of schizophrenic patients not only to move among their delusional worlds and the 'real' world of consensus but also to register their ontological differences.[44] One of his patients described this capacity to maintain a mental account of this crucial difference as 'double bookkeeping'. Louis Sass observes:

> It is remarkable to what extent even the most disturbed schizophrenics may retain, even at the height of their psychotic periods, a quite accurate sense of what would generally be considered to be their objective or actual circumstances. Rather than mistaking the imaginary for the real, they often seem to live in two parallel but separate worlds: consensual reality and the realm of their hallucinations and delusions.[45]

This 'double bookkeeping' describes well the ways that Bone moves between distinct ontological orders and reflexively notes their differences, including the different Nettas that he experiences.

First-person accounts of schizophrenia often include such explicit reflections on the relations among the phenomenologically distinct worlds in which patients find themselves. In her memoir of schizophrenia, *Operators and Things*, O'Brien writes:

> Caught up in my new world, and with the world of sanity almost wiped from my mind, the resemblance between the two worlds was not apparent until afterward, when I walked into a psychoanalyst's office at the advice of my voices, and gave him the message they had told me to give him.[46]

While remarking upon the feelings of unreality, depersonalization, and instrumentalization that defined the phenomenology of her delusional world, O'Brien also observes points of resemblance between that world and the 'world of sanity' as defined by her psychoanalyst. This similarity suggests a relation not of reducibility but rather of resemblance and adaptation from one world to another.

O'Brien appears to make this point when she notes that her feeling of being reduced to a 'Thing' that is manipulated by 'Operators' resembles the evolving, psychiatric view of the mind to which she had been subjected.

> It was a somewhat startling discovery to me. I had always thought of the unconscious as a whirling pool of repressed emotions, better repressed. Instead, it appeared to be a sort of private Univac, an incredible piece of thinking mechanism [. . .][47]

Whereas Jung's patient felt that her mind had been reduced to a telephone exchange, O'Brien experiences her mind as an early mainframe computer that passively carries out the requests made upon her by malevolent operators. O'Brien indicates that she has adapted her own technological delusion from the discourse of her doctors while marking certain important exceptions:

> to a psychiatrist the unconscious mind, in insanity, is a machine out of working order, no more, no less. But this one psychiatrist apparently had come upon evidences of purposefulness in the schizophrenic's mind. And he had veered from the psychiatrist's usual assumption that the schizophrenic finds himself in an uncontrolled, bizarre and crazy dream to wonder if the dream world might have been created on an unconscious level with intention and purpose.[48]

O'Brien marks the difference, first elaborated by Karl Jaspers, between explanations via neurological or informatic causes and a phenomenological understanding of the patient's delusional world. Jaspers writes, '[c]ausal thinking impinges on what is alien, non-understandable and on what can be manipulated; understanding of meaning impinges on myself in the other and on what is closest to me in the other.'[49] Like Bone in *Hangover Square*, O'Brien describes a movement between these two epistemes. Her memoir demonstrates the ways in which the technological worlds of schizophrenic delusions often reflect the mechanistic forms of explanation that have recurred across the history of psychiatry. More recent work in psychiatry and cognitive neuroscience has conformed to O'Brien's intuition that the brain is often regarded as a computer. Christopher Frith observes:

> This attitude toward the mind-brain problem is much influenced by experience with computers. We can describe the operation of a computer in terms of processes within and interactions between central processors, memories, and response buffers. Alternatively, we can describe these processes in terms of electronics. We can happily cross the boundary between these two levels of description when we say, 'here is the memory chip'.[50]

In many ways, little has changed since Bleuler suggested that the mind 'works like an electric device where functions are switched on and off.' In place of the telephone exchange, the computer has emerged as the device on which the mind is modeled in both clinical accounts and psychotic delusions. Peters writes that

> each age gets the form of madness it deserves and that every form of madness is a parody of the reigning form of reason. Pathology reveals

> normality. In the same way, each format or technology of communication implies its own disorders.[51]

The content and structure of psychotic delusions, rather than being beyond the scope of understanding, often manifest distorted reflections of the scientific explanations of the mind's functions and dysfunctions.

If psychotic experiences were once regarded as the senseless epiphenomena of neurological dysfunctions, or, as Karl Jaspers put it, 'un-understandable', an alternative approach would consider the ways that delusional formations constitute worlds in which experiences of voice-hearing, hallucination, and thought broadcasting might be explained – worlds that are as culture-bound and historically determined as any work of literary fiction. In this way, it is possible to recognize how Bone's delusion is hardly unique but is only one example of a common adaptation of the experiences of radio, telephone, and film to explain his own hallucinatory and delusional states – a work of a narrative world construction.

In *Ways of Worldmaking*, philosopher Nelson Goodman inquires into the ways that worlds may appear to come into being and from what they are fashioned:

> Not from nothing, after all, but from other worlds. Worldmaking as we know it always starts from worlds already on hand; the making is a remaking. [. . .] My interest here is rather with the processes involved in building a world out of others.[52]

Adapted to the study of narratives of psychopathology, Goodman's observation helps us to recognize the ways that delusional worlds are constructed from the culturally available explanatory frameworks. A proper understanding of his world would therefore demand not only a sense of its internal phenomenological structure but also its imbrication in existing explanatory frameworks as well as a sense in which his world is the product of an act of narrative fashioning.

Here, it is worth recalling Jaspers's critique of psychiatry's tendency to disregard the phenomenological worlds of its patients – the tendency to treat their worlds as simply not real – as well as his suggestion that '[t]he methods for knowledge in this field [of psychopathology] are only acquired through a training in the humanities.'[53] I have taken Jaspers's suggestion to mean, in part, that the study of unconventional fictional worlds, such as *Hangover Square*, may lead toward a better understanding of delusional worlds and vice versa. Fiction may provide training for accessing other narrative worlds with other kinds of ontological status – including memoirs of mental illness. It is in this way that the study of experimental narratives may be enlisted to answer Jaspers's call for an approach to mental illness that is informed by the methods of the humanities and that strives not only for explanation but also for understanding.

In her study of voice-hearing, Lisa Blackman argues for an approach to mental illness in which 'we would no longer be interested in mapping underlying internal psychological or neurological mechanisms, and would instead concentrate on the different ways in which the voice-hearing experience is mediated through culturally available narratives.'[54] It is this approach that has guided this discussion of *Hangover Square*, *Operators and Things*, and other works of modern fiction and memoirs of mental illness. Reconnecting narratives of voice-hearing, thought broadcasting, and depersonalization to the network of available explanatory frameworks available to the subjects in question brings these experiences and delusional accounts into legibility. Angela Woods writes,

> recognising that stories of illness emerge as artefacts of a shared social world in no way detracts from the potential of those narratives to illuminate an individual's experience; on the contrary, it goes some way to explaining why illness narratives effect personal and social transformation.[55]

To this end, we might regard the recurrent technological and depersonalized qualities of voice-hearing in schizophrenia not merely as singular or senseless effects of dysfunctional brains but as reflections of available explanatory discourses. Such an approach, as I've modeled here, would not only bring into focus the ways that narratives of mental illness manifest altered phenomenologies but would also account for the recurrent, technological metaphors that subjects often deploy in order to explain voices and worlds that have often been regarded by psychiatrists as 'ununderstandable'. Writing in critique of the long history of abuses psychiatric patients have suffered, Richard Bentall writes, '[a]bove all, without regarding psychosis as *ununderstandable* (to use Jaspers' terminology), it would not have been possible to deny psychiatric patients a voice.'[56] Perhaps, by reconnecting the experiences and writings of the mentally ill to their cultural, technological, and phenomenological worlds, it might be possible to make their voices heard.

Notes

1 Friedrich A. Kittler, *Discourse Networks, 1800/1900*, trans. Michael Metteer, with Chris Cullens, foreword David E. Wellbery (Stanford: Stanford University Press, 1990 [1985]).

2 For a discussion of the novel as an addiction narrative, see Geoff Ward, 'The Wibberlee Wobberlee Walk: Lowry, Hamilton, Kavan and the Addictions

of 1940s Fiction', in *The Fiction of the 1940s: Stories of Survival*, ed. Rod Mengham and N. H. Reeve (New York: Palgrave, 2001), pp. 26–45.

3 Patrick Hamilton, *Hangover Square; A Story of Darkest Earl's Court in 1939* (New York: Europa Editions, 2009 [1941]), p. 71.

4 Ibid., p. 72.

5 'The London-Paris Telephone Open Today', *Daily News* (1 April 1891), 2 (cit. in David Trotter, *Literature in the First Media Age: Britain Between the Wars* (Cambridge, MA: Harvard University Press, 2013), p. 48).

6 Trotter, *Literature in the First Media Age*, p. 47.

7 Hamilton, *Hangover Square*, p. 99.

8 Ibid., p. 31.

9 Ibid., p. 30.

10 Ibid., p. 257.

11 Ibid., p. 47.

12 C. G. Jung, *The Psychology of Dementia Praecox*, trans. R. F. C. Hull (Princeton: Princeton University Press, 2015 [1974]), p. 149.

13 Ibid.

14 Avital Ronell, *The Telephone Book: Technology, Schizophrenia, Electric Speech* (Lincoln: University of Nebraska Press, 1989), p. 132.

15 E. Kraepelin, *Psychiatrie: Ein Lehrbuch für Studirende und Aertze*, sixth fully revised ed., 2 vols (Leipzig: Johann Ambrosius Barth, 1899), pp. 173–4 (quoted in John Durham Peters, 'Broadcasting and Schizophrenia', *Media, Culture & Society*, 32.1 (2010), 123–40 (pp. 131–2)).

16 Daniel Paul Schreber, *Memoirs of My Nervous Illness*, trans. Ida Macalpine and Richard Hunter (New York: New York Review of Books Classics, 2000 [1903]), p. 21.

17 Barbara O'Brien, *Operators and Things: The Inner Life of a Schizophrenic* (Cambridge: Arlington Books, 1958), pp. 2–3.

18 Hamilton, *Hangover Square*, p. 28.

19 Eugen Bleuler, *Naturgeschichte der Seele und irhes beisstwerdens, eine Elementarpsychologie* (1921) (quoted in Verena Kuni, '"Je Suis Radio" – From the "Influencing Machine" to the "Radio- Self"', in *The Air Loom and Other Dangerous Influencing Machines*, ed. Thomas Roske and Bettina Brand-Claussen (Heidelberg: Wunderhorn, 2006), pp. 239–53 (243)).

20 Kuni, '"Je Suis Radio"', p. 245.

21 C. Moffett, 'An Interview with Professor Alexander Graham Bell', *McClure's Magazine*, 1 (1893), 39–43 (quoted in Peters, 'Broadcasting and Schizophrenia', p. 130).

22 Peters, 'Broadcasting and Schizophrenia', p. 129.

23 Darian Leader, *What Is Madness?* (New York: Penguin, 2012), p. 73.

24 See Andrew Gaedtke, *Modernism and the Machinery of Madness: Psychosis, Technology, and Narrative Worlds* (Cambridge: Cambridge University Press, 2017).

25 Wyndham Lewis, *The Art of Being Ruled*, ed. Reid Way Dasenbrock (Santa Rosa: Black Sparrow Press, 1989 [1926]), pp. 105–6.

26 Ford Madox Ford, *Parade's End* (New York: Penguin, 2001 [1924–28]), p. 281.

27 Ibid., p. 282.

28 For more on the telephonic metaphor for the shell-shocked mind, see Heather Fielding, *Novel Theory and Technology in Modern Britain* (Cambridge: Cambridge University Press, 2018), pp. 59–85.

29 Ford, *Parade's End*, p. 270.

30 Ibid., p. 652.

31 Steven Connor, *Dream Machines* (London: Open Humanities Press, 2017), pp. 8–24.

32 Marie Corelli, *The Life Everlasting: A Romance of Reality* (New York: A. L. Burt, 1911), pp. 373–4 (quoted by Connor, *Dream Machines*, p. 121).

33 Louis Sass, *The Paradoxes of Delusion: Wittgenstein, Schreber, and the Schizophrenic Mind* (Ithaca: Cornell University Press, 1995), p. 22.

34 Karl Jaspers, *General Psychopathology*, trans. J. Hoenig and Marian W. Hamilton (Baltimore: Johns Hopkins University Press, 1997), p. 704.

35 Hamilton, *Hangover Square*, p. 24.

36 Ibid., p. 27.

37 Ibid., p. 17.

38 Ibid., p. 99.

39 Daniel Hutto, 'Narrative and Understanding Persons', in *Narrative and Understanding Persons*, ed. Daniel Hutto (Cambridge: Cambridge University Press, 2007), pp. 1–16 (3).

40 Hamilton, *Hangover Square*, p. 24 (my emphasis).

41 Jerome Bruner, 'The Narrative Construction of Reality', *Critical Inquiry*, 18.1 (1991), 1–21 (7).

42 Hamilton, *Hangover Square*, pp. 326–7.

43 Ibid., p. 327.

44 See Eugen Bleuler, *Dementia Praecox; or, the Group of Schizophrenias*, trans. Joseph Zinkin (New York: International University Press, 1964), pp. 65, 127–30.

45 Sass, *The Paradoxes of Delusion*, p. 21.

46 O'Brien, *Operators and Things*, p. 8.

47 Ibid., p. 135.

48 Ibid., p. 126.

49 Jaspers, *General Psychopathology*, p. 462.

50 Christopher Frith, *The Cognitive Neuropsychology of Schizophrenia* (Hove: Erlbaum Associates, 1992), p. 25–6.

51 Peters, 'Broadcasting and Schizophrenia', p. 124.

52 Nelson Goodman, *Ways of Worldmaking* (Indianapolis: Hackett, 1978), pp. 6–7.

53 Jaspers, *General Psychopathology*, p. 293.

54 Lisa Blackman, *Hearing Voices: Embodiment and Experience* (London: Free Association Books, 2001), p. 77.

55 Angela Woods, 'Post-narrative: An Appeal', *Narrative Inquiry*, 21.2 (2011), 399–406 (403).

56 Richard Bentall, *Madness Explained: Psychosis and Human Nature* (New York: Penguin, 2003), p. 498.

5 'The Zoom of a Hornet'

Virginia Woolf, Aural Biopolitics, and the Phenomenology of an Air Raid

Beryl Pong

On 1 September 1940, on the eve of the Blitz and before a period of nightly and severe air-raiding in London began, a memo was drafted and sent to Mass Observation:

Memo on the Senses re. Air Raids.
Touch. This has not yet acquired great specific application.
Where bombs fall the physical effects of Blast etc on the mind, and the anticipatory idea of wounds might be gauged as against Sounds & Sights.
[. . .]
Smell (& Taste) Inv[estigator]: doesnt know that this enters into the pictures at all as yet, but in the event of Gas being used it certainly will.
Sound. Sirens, Bombs, Gunfire, near and far, the Stuttering note of German planes, etc.
Though there is so much in the way of Sounds, Inv: suggests that –
Sight Is the most important medium for reactions.
In spite of the vastly increased danger the need to look seems to amount almost to a compulsion with many people.[1]

Much can be said about the memo, not least its preoccupation with the visual and the 'compulsion' to look, as well as the olfactory memory invoked from past wartimes. I linger here, however, on the suggestively abortive observation of the aural and on what this might tell us about the phenomenological experience of wartime listening. The description is incomplete, dashed, and curtailed, merely, 'there is so much in the way of Sounds' that is difficult to pin down, suggesting both the interruption and truncation of thought, as well as an unarticulated 'more' to the soundscape which lies beyond what is delineated. There is a sense that one hears both not enough and too much.

What is also striking about the memo is not only its depiction of those elements that 'accumulate upon the senses', but its premonitory stance: its

DOI: 10.4324/9780429291616-7

anxiety about the sensuous injuries that an air raid would inflict *in the future* and how that expectation seeps into the present of pre-wounding. Written just before the Phoney War ended, the memo is preoccupied with the impact of violence yet to be experienced – with 'the anticipatory idea of wounds'.[2] This chapter is concerned with the particular act of 'listening to anticipation', which did not dissipate with the outbreak of war. As civilians came to expect carnage on a regular basis, they not only faced the question of if but when and where bombing would take place. Literary fiction indexed how anticipating injury became a quotidian reality in a more traumatic way.

The complex interrelations between listening *in*, *to*, and *for* wartime and their psychic effects are foregrounded in Virginia Woolf's 'Thoughts on Peace in an Air Raid' (1940). In it, she describes an auditory environment in which it is hard to think with temporal continuity or to think self-consciously, much less politically or ethically. 'It is a queer experience,' she writes, 'lying in the dark and listening to the zoom of a hornet which may at any moment sting you to death. It is a sound that interrupts cool and consecutive thinking about peace.'[3] Her thinking, and her struggle to think, presages what Geoffrey Bennington would theorize as the aporia of emergency:

> what is urgent is to hold up against urgency so as to think it, to be patient enough with the emergency to do its urgency justice by not just running with it: the thinking of urgency will, then, also be a *thinking* of the resistance to urgency.[4]

What emerges from Bennington's writing and its not-quite-running-away syntax is the struggle to take urgency at face value, to not diminish it, while questioning and refusing its grounds and logic: a double thought-endeavour required to restore the ethical and political dimensions of a consuming experience like aerial bombardment, whose routinized nature makes the immediate future seem foregone. For Woolf, the resistance to urgency would begin earlier, in anticipation, with her 1937 novel *The Years*. There, she rendered this aporia of emergency by creating a literary air-raid siren of sorts, conjuring the soundscape of the war metropolis while warning against the lure – the siren-call – of succumbing to its unthinking urgency.

Listening to Anticipation

Much recent work in war and sound studies has focused on the more dramatic noises of weaponry and destruction, bombing and shelling.[5] But we need to pay attention to the sounds of everyday life as well. 'A particular idea of what war *is* has produced a narrow definition of wartime sound and audition,' Jim Sykes observes, 'the result being the exclusion of many uses of sound and listening that would emerge as integral to war if one were to adopt the broader notion of "wartime violence"'.[6] In his concept of 'wartime acoustemology' – a sonic way of knowing and being in the wartime world – Sykes includes the more mundane soundscapes with which wartime sounds are intermingled, such as the silence of suspended transport and those of shuttered schools and shops.[7]

This broader idea of wartime acoustemology is worth bearing in mind for the Blitz. Although the home front endured nearly seventy-six consecutive nights of air raids, as we will see from medical reports and diaries from that time, the sounds of suffering and devastation were punctuated by quieter moments of boredom, respite, and silent fear. If we view war not only in terms of its 'loud' instantiations but as 'a derangement, rearrangement and resignification of everyday life' more generally, then wartime can encompass a whole auditory environment that demands distinctive and sustained kinds of listening in less obvious ways.[8] What Steven Connor calls the 'grotesquely widened bifurcation of visuality and hearing' of an air raid, for instance, the way bombs and bombers target on sight while the targeted can only rely on sound, might also describe quieter, more routine measures like the blackout, when one listened more to compensate for seeing less.[9]

The British home front is also an interesting case study for wartime acoustemology, because the altered experience of everyday life began before air raids, even before the war's official outbreak. Some accounts of the Second

Figure 5.1 In the road safety poster, 'Lookout in the Blackout – Think Before You Cross the Road', by Pat Keeley, c. 1940, the ear helps one to 'look' in the blackout.

Source: © The National Archives.

World War might peg its beginning to the so-called regional conflicts of the Spanish Civil War (1936–39) or the Second Sino-Japanese War (1937–45) rather than to the German invasion of Poland. There existed, even before the Munich agreement, a long build-up of war anxiety, which centred on the expectation of civilian aerial bombardment especially.[10] In this way, we can contextualize the psychology of anxious listening within the 'anticipatory idea of wounds' that characterized much of the mid- to late 1930s. Like the Mass Observation memo, Lewis Mumford's *The Culture of Cities*, published in 1938, describes a cacophonous war metropolis before the fact:

> The sirens sound. Schoolchildren, factory hands, housewives, office workers, one and all don their gas masks. Whirring planes overhead lay down a blanket of protective smoke. Cellars open to receive their refugees. Red Cross stations to succor the stricken and the wounded are opened at improvised shelters . . . Presently the anti-aircraft guns sputter. Fear vomits: poison crawls through the pores. Whether the attack is arranged or real, it produces similar psychological effects [. . .] Panting, choking, spluttering, cringing, hating, the dweller in Megalopolis dies, by anticipation, a thousand deaths. Fear is thus fixed into routine: the constant anxiety over war produces by itself a collective psychosis comparable to that which active warfare might develop.[11]

This passage is notable not just for its sensationalist rendering of urban destruction but for its insistence on emergency as protocol. What wounds is not just the catastrophic event but mobilization: 'the series of protocols – sirens, searchlights, gas masks, scrambling, anti-aircraft guns, engines, bombs – whose terminus is the catastrophe', Paul Saint-Amour writes.[12] The depiction of the state of emergency as 'routine' is perhaps striking to us but would not have been to Mumford's contemporary readers, for whom the anticipation of war had already warped the experience of urban life. Crucially, Mumford blurs and purportedly finds irrelevant any differences between wartime and peacetime by refraining from stating whether the raid is actual or a rehearsal. The habits of mobilization mean that time itself, '[n]o longer a passive field within which violence unfolds – and anticipation in particular – has become a new medium for delivering injury', Saint-Amour adds.[13] An injurious anticipation, then, is *the* affect of the state of emergency; one 'dies, by anticipation, a thousand deaths.'[14]

The affect of injurious anticipation is linked to sound. Mumford's prose, although exaggerated and melodramatic, renders the war metropolis as a confusing, unrelenting, visceral soundscape, where the noises of mobilization entwine with those of the cityscape. And what Mumford anticipates did partly materialize as numerous medical studies of the day testified to the harmful, enervating effects of hearing in war. '[F]ear seems to be linked above all with *noise*,' Mass Observation concluded from its enormous social survey.[15] 'It is the siren or the whistle or the explosion or the drone – these are the things that terrify . . . Fear seems to come to us most of all through our sense of hearing.'[16] Psychologist Edward Glover, in his 1942 study of war neuroses, noted that civilians suffered from the pathological inability to interpret sound even as they had an 'obsessive concern' with doing so.[17] In his article 'Psychological Effects of Air-Raids' (1941), P. E. Vernon also observed this difficulty in interpreting sound, particularly in terms of how to distinguish between sound's wartime referents versus its everyday manifestations: 'The mere sounding of the sirens disrupted a sense of normality [. . .] Their gruesome wail readily frightened people, and, at nights, the noises of automobile or tramcar gears were continually mistaken for warnings.'[18]

Life-writings and diary entries repeatedly show how aerial bombardment confounds while provoking acts of auditory orientation. On 10 December 1940, the writer Sylvia Townsend Warner wrote: 'A long distant growl. An explosion? Not until the third of these did we realise it was only a thunderstorm. One of the advantages of a thunderstorm is that one feels no need to locate it.'[19] Later in 1944, after the advent of the V-1 and V-2 rockets, a civilian named Vivienne Hall noted in her diary:

> And so it goes on – day and night these flying bombs come . . . droning across the skies, sometimes hourly, sometimes at 10 minute intervals, sometimes several at once, in fact there is no knowing which of the systems is being used so you just have to listen and hope for the best! That's all we do, we are just one large frightened ear – each sound of traffic, each purr of an engine stops conversation whilst we sort out the sound and decide whether to take cover or not . . .[20]

Always on the lookout – or hear-out – for the bomb that would hit home, being in the war metropolis meant being an attentive, fearful, nervous, and discerning auditor. As the sounds of war become confused with the banal, war morphs from event into climate.[21] These documents speak to what Mary Favret identifies as 'the troubling but enduring marriage of the everyday with war . . . [which] ought to be read as a symptom of the history of modern warfare.'[22] It is a marriage in which the everyday seems to sustain 'under the veneer of peace' the work of war both before war's beginning and after its end – where the very experiences of peacetime and wartime seem continuous rather than discontinuous from the other.[23]

The Aural Biopolitics of the Air-Raid Siren

The aforementioned diary entries focus on private sensations of listening to war, but sound is spatial – it takes place in space – and it has social dimensions which cannot be ignored. This mattered during the Blitz because of the pervasive worry that an aerial attack would unleash mass hysteria and panic. No less than 4 million incidents of psychological debilitation were expected in the first six months of a second global war, where mental cases were thought to outnumber physical cases by 3 to 1.[24] In 1938 (again, in anticipation), a committee of psychiatric experts had written a report detailing what they perceived would be the most significant problems of war, which included the possibility of an 'epidemic of shell-shock' similar 'to that observed during the war of 1914–1918'.[25] Individual well-being was seen as inextricably tied to the well-being of the collective – as a public health concern.

Because of this, acclimatization to Blitz sounds – necessary to acclimatize to the climate of war – 'was constructed as a moral duty of the patriotic wartime citizen'.[26] In *The Age of Noise in Britain* (2016), James Mansell discusses how this acclimatization could be cultivated by various means, such as 'the adoption of psychological "nerve training"'.[27] The improbably named self-help book *Nerves Versus Nazis* (1940), for instance, by the writer and journalist John Langdon-Davies, suggested that resilience to wartime noise is a contribution to the war effort. 'This is not only a war of machines, it is a war of nerves,' he writes.[28] 'NOISE IS THE GREATEST ENEMY.'[29]

> By perpetual bombing people can be kept awake, kept worried, kept restless, kept unoccupied, and this is exactly what is wanted. Noise is the way to do it . . . It is the bark of a bomb that is worse than its bite.[30]

Because acclimatization was seen as a choice, Mansell writes, the fear of noise 'separated bad citizens from good citizens, those who counted as fully national and those who threatened to destabilize the nation by failing to "hold their nerve"'.[31]

Borrowing from William Reddy's theory of 'emotional regimes' – cultures that determine how freely one may articulate, own, and work through their emotions – Amy Bell suggests that the emotional regime of wartime London's soundscape 'emphasized civilian bravery as the road to national survival; it was one in which fears and anxieties could not be overtly expressed'.[32] While in one sense a means of promoting resistance, resilience, and psychological fortitude, wartime's emotional regime, especially surrounding sound, also downplayed – indeed concealed – the actual violence of war. It asserted that civilians were responsible for how they acted and felt in wartime despite war being a condition to which they were subjected without choice.

Wartime listening is therefore conditioned by an aural biopolitics. As Michel Foucault writes in *Society Must Be Defended* (1975–76): 'Biopolitics deals with the population, with the population as a political problem, as a problem that is at once scientific and political, as a biological problem and as power's problem.'[33] Certainly, different understandings of biopolitics and biopower have emerged in contemporary times, though for our purposes here I remain concerned with Foucault's understanding of biopolitics as the management of human bodies and processes under regimes of authority over knowledge, power, and subjectivation.[34] Because of the wartime context, I also remain concerned with how biopolitics legitimizes the act of state intervention upon individual lives while, at the same time, encouraging them to construct and interpellate themselves in relation to the 'health' of the collective.

Biopolitics, for Foucault, is about managing and administering life – and, in wartime, there was arguably no device more effective at interpellating the individual, at calling upon an extant anticipation of violence, than the air-raid siren. ('The sirens sound,' Mumford writes.[35]) Encouraging almost instantaneous reaction without demanding reflection, the siren 'broadcasts distress', as Steve Goodman puts it in *Sonic Warfare* (2010). 'It is a centrifugal sound designed to scatter people in its path by pulsing waves of nonlinguistic command to disperse a population.'[36] 'A siren obviously signifies alarm, but [. . .] its very modulation of frequency produces a state of alert that can undermine and override cognition,' he notes. Thus, sirens are 'a whole directly affective signifying semiotics of emergency, a call to action, the inducement of a state of readiness, initiating a kind of technical antiphony.'[37] In keeping with *Nerves Versus Nazis*, the air-raid siren suggested action rather than reaction, calmness rather than hysteria, preparation rather than helplessness. It is biopolitical in making the problem of emergency containable, manageable, both for the individual body and for the body of the population.

Where earlier experiments for warning sirens assumed that enhanced attentiveness (which the risk of air attack would create) and the prohibition of the use of sirens for anything other than warning purposes were enough to signal emergency, later tests determined that changes in pitch were more likely to attract attention. This led to the siren's distinctive octaval quaver, which produced much commentary. In the November 1940 issue of *Musical Times*, three contributors attempted to describe the 'musical notes' of sirens, with one observing that the 'alert signal has time as well as pitch':

> The E sharp and G sharp are set down as a semiquaver because the average siren seems to touch on them for only an instant before moving upwards [. . .] During the pause on D and B there is usually a good deal of G in the air.[38]

(See Figure 5.3.) Despite the glib tone, the writer makes a remarkable observation: because the Warning signal warbles between two notes while the All-Clear has one, pauses on any note 'suggested to some people that an all-clear was being sounded [. . .] [therefore] some sirenists have lately

Figure 5.2 An ARP siren at Snow Hill Police Station, in the City of London
Source: Photo by © Hulton-Deutsch Collection/CORBIS/Corbis via Getty Images.

taken to cutting out the pause on the top notes.'[39] With the Warning and the All-Clear differentiated primarily by duration and by tonal shift, the temporality of the siren becomes a microcosm of the temporality of an air raid itself, where a conditional and inevitable future is inextricable from a prolonged, suspended present: where the end might not be the end at all but merely the illusion or misapprehension of an end. The siren, in Ryan Bishop and John Phillips's words, represents 'both life and death, before and after, prefacing in a dire prolepsis the post-apocalyptic event before the event'.[40] While cultivating nerves, the siren frays them as well. As one wartime diarist put it, 'I can't bear it. I can't *bear* it. If them sirens go again tonight, I shall die. It's me nerves, they're all used up.'[41] The siren is both the response *to* fear and the object of fear.

British literature of the war variously represented such mixed, fraught meanings underpinning individual trauma and civic duty. In Graham Greene's comic short story 'Alas, Poor Maling' (1940), the titular character has borborygmi, a harmless form of indigestion that causes his stomach to have 'an ear', and, absurdly, to record and replay external sounds: it 'used to pick up notes in an extraordinary way and give them out

Air-Raid Sirens

The alert signal has time as well as pitch. Most of the sirens to which I have listened in my two districts (central London and a south-west suburb) have sung their *leitmotif* more or less as follows (all the movement being, of course, *portamento*) :

The first bar may last for five to seven crotchets. The semibreve in bar 3 is probably too long. The E sharp and G sharp are set down as a semiquaver because the average siren seems to touch on them for only an instant before moving upwards; I have rarely heard a siren move down to G and E and hold them. The ascent is noted in quavers because it is decidedly quicker than the descent.

During the pause on D and B there is usually a good deal of G in the air.

When a number of alerts are audible at the same time the long-held D and B tend to dominate the chorus, and may have suggested to some people that an all-clear was being sounded. Perhaps it is for this reason that some sirenists have lately taken to cutting out the pause on the top notes.

W. McNaught.

Figure 5.3 From W. McNaught, A. Langmead Robinson and E. H. Walker, 'Air-Raid Sirens', *Musical Times*, 81.1173 (November 1940), p. 457

Source: Reproduced by permission of the *Musical Times*.

again'.[42] In addition to 'The Lambeth Walk', his gestational gramophone picks up and broadcasts the air-raid siren, interrupting a major commercial merger (by businessmen hoping to avoid new wartime income-tax regulations) and causing it to nearly fall through. While dealing playfully with the idiom that nothing is certain except death and taxes, Greene's story suggests both the untimeliness and extended temporality of states of emergency: 'poor Maling's stomach,' the story ends, 'had picked up the note of the Warning only too effectively, but it had somehow never taken to the All Clear.'[43] The story is frankly ridiculous, but its improbability is attenuated by the fact that it pointedly takes place late in the Phoney war – on 3 September 1940, the anniversary of war's outbreak, when a mass attack was expected – and by the fact that diarists often located the siren's physiological symptoms somatically, in the stomach.

As diarist Marie Lawrence wrote: '26 August 1940. 1 a.m. Still Jerries were over. The air was alive with them. It was just like having a bad stomach ache. Slowly the pain would come on, get worse and then die down.'[44] Hence the many wartime advertisements for laxatives, digestives, and diuretics. For example, the *Daily Mirror* advertisement for Kruschen Salts (9 September 1940), a laxative-diuretic, claims to relieve the psychological jitteriness of wartime. In response to questions probing wartime mental and physiological health – 'Do you jump when a bus backfires?' 'Does worry bring sleepless nights?' – the advertisement proposes Kruschen Salts as the remedy that will 'keep your bowels open and your kidneys well flushed', since '[t]o do your bit you must be fit'.[45]

Air raids and air-raid sirens also underpin a prominent literary trope of the day: amnesia or memory loss. In Margery Allingham's crime novel *The Traitor's Purse* (1940), amnesic crime detective Albert Campion trips over an alarm that sounds like an air-raid siren, and his body responds with terror: 'It screamed at him, sending every nerve in his body tingling to the roots of his hair. It bellowed. It raved. It shrieked, trembling hysterical into the night [. . .]'[46] Air-raid sirens similarly fill another detective novel, Greene's *The Ministry of Fear* (1943), which dramatizes the impulse to use sound to locate space. Here, the main character, who is attempting to unravel an espionage plot, uses the telephone to ring up those he knows, and it is only when the phone is answered that he can be assured that 'that part of the world was the same'.[47] This character also forgets the past nineteen years of his life from a bomb blast, and, again, the interest in memory loss has some historical context: a *British Medical Journal* article of September 1939 suggested that the expected civilian response to aerial bombardment included concussion, stupor, and coma, as well as forms of somnambulism and amnesia, although the reality was less extreme.[48]

In wartime writing, the trope of memory loss powerfully encapsulates an air raid's ability to reduce the experience of time to a present seemingly divorced from history or continuity. And because these are detective-crime novels or spy thrillers about fascism and fifth columnists, the authors suggest that this kind of temporality vexes one's capacity to act self-consciously when it comes to matters of ethics or politics. But if the function of the siren is to alert, and if the affect of emergency is anticipation, we need to consider earlier literature, the literature of forewarning, and how such literature sought to marshal against both the aural biopolitics of mobilization and the sonic violence of the war to come.

Between Memory, Rehearsal, and Prolepsis: The Soundscape of *The Years*

Woolf's *The Years* might seem like an unlikely candidate as a siren for a war that has not yet happened. Conventionally seen as a return to the

Victorian-realist novel, the failure of which led to a radically different vision with her final, unfinished work, *Between the Acts* (1941), the text unravels without much sense of momentum or climax. Tracing three generations of the Pargiter family, its chapters hop unevenly forward from 1880 to the 'Present Day', and the details of everyday life are foregrounded over world-historical events, such as the lightly registered deaths of Parnell and Queen Victoria.[49] But as Woolf's diaries and letters remind us, the lengthy composition of *The Years* was intertwined with her apprehensions concerning a very contemporary political and historical development: the rise of fascism and the inevitability of a Second World War.[50] The novel's simultaneous continuity and discontinuity – time as both eventful and uneventful, as chronological yet staccatoed and disjointed, and even synchronous, as phrases and objects repeat throughout – seems to both gesture towards and prepare for the temporal dislocations that a new air war will bring.

A mainstay of Woolf's oeuvre is the relationship between the personal and the societal, the atomistic and the whole, which also underpins the individualizing and collectivizing work of biopower. This theme is especially pronounced in her late works of the thirties and forties, and although culminating in the 'Unity – Disperity' refrain of *Between the Acts*, it is a structuring principle of *The Years*, not least on the level of narrative form.[51] Woolf originally envisioned the book as a hybrid form that would alternate fictional 'extracts' or chapters from a book of explanatory essays of social critique, but she eventually dropped the essay portion for fear of proselytizing and propagandizing. In Maud Ellmann's summation, 'she pruned *The Years* of anything that could be taken for grandstanding'.[52] As Woolf would write in 'The Leaning Tower' (1940), she opposed 'the pedagogic, the didactic, the loud speaker strain' of the Auden generation, whom she accused of protesting too much.[53] Because her attack on the 'loudspeaker' also referred to the loudspeakers of the Nazi and fascist meetings of the 1930s and to Hitler's voice that would be heard on the radio, her work suggestively links authoritarianism to literary monologism and univocality.[54] Her opposition to explicit didacticism eventually resulted, in *The Years*, in doubled perspectives: each chapter begins with a brief, omniscient, autocratic vision of London, followed by a longer, sustained narrative about the details of everyday life that the former does not allow.

Critics have focused on the novel's pointed aerial views in effecting this tension between the individual and the collective.[55] But in opposing 'the loud speaker strain', *The Years* not only undermines visual but auditory omniscience as well – it opposes not only panoptics but what Goodman calls 'pansonics'.[56] For this is a very noisy novel in which the sounds of characters' voices mingle with those of street hawkers and barrel-organs, with overheard songs as well as opera music, with the cooing of birds, the ringing of church bells, the clacking of horse hooves, the running of car engines, and the popping of guns and the sounds of sirens from the

First World War. These sounds are often presented to us without narrative hierarchization, heard by a specific character rather than an omniscient auditor. The effect is something akin to an aural representation of lived geography.

For example, after the stilted domestic environment of the opening '1880' chapter and its death-bed scene, the '1891' chapter follows Eleanor Pargiter as she goes out for a walk on London's streets. Emerging from her Victorian home, Eleanor feels a sense of openness as the urban noise swirls around her: '[S]he heard the dull London roar with pleasure. [. . .] This was her world; here she was in her element.'[57] 'The roar, the confusion, the space of the Strand came to her with a shock of relief. She felt herself expand.'[58] We know from the three scrapbooks that Woolf compiled between 1931 and 1937 (which fed into her 'Novel-Essay', *The Years* and *Three Guineas*, though she saw them as 'one book') that she had feminist sociological ambitions: many of the clippings she culled related to women's education, employment, and family life.[59] The voices around her, the sound of traffic, the paperboys crying, the pigeons cooing: the urban cacophony gives Eleanor a sense of freedom and agency.

By the middle of the novel, however, with the impending outbreak of the First World War in the '1914' chapter, sound seems to index a very different kind of personal and social experience. The atmosphere is premonitory and foreboding: 'The air over London seemed a rough sea of sound through which circles travelled. But the clocks were irregular as if the saints themselves were divided. There were pauses, silences [. . .]'.[60] Martin and Sara's journey from St. Paul's to Hyde Park is characterized by moments of 'harmony' and calm, but also, by the narrative's insistence on the 'ripple in the air [. . .] all the clocks of the city [. . .] seemed to be gathering their forces together; they seemed to be whirring a preliminary warning'.[61] Rather than absorbing or fluidly merging with the sounds around them, the characters' thoughts and speech are constantly interrupted and thwarted. 'Fragments of other people's talk reached them in broken sentences'; 'it was impossible to talk'; and meanings are missed 'in the roar of the traffic'.[62] Hints of political tension from contemporary events are conveyed through sonic intrusion: 'Roll up the map of Europe . . . I don't believe in force!' says a diner in the restaurant.[63] Casual xenophobia is broached this way, too. '"Ladies and gentleman . . ."', a streetside orator calls, '"Don't be afraid [. . .] Do I look like a Jew?"'[64] Instead of personal agency or emancipation, the '1914' chapter uses sound to insist on the porosity between the private, the social, and the geopolitical, and on a certain hold the latter has over the former.

Anna Snaith has noted a fascination in this novel with the role of music, especially with how particular appearances of nursery rhymes, popular waltzes, and classical music relate to 'the promotion of nationalism'.[65] In her critical edition, she also argues that *The Years* is fundamentally concerned with the regulation of domestic and national space by sound, and with how private and public spaces are constantly subject to sonic intrusion. She suggests that the novel is 'laid out through aural markers' as much as through physical landmarks, and this is corroborated by Elicia Clements, who remarks that the text creates a kind of 'aural architecture' in its commitment to representing the 'properties of a space [. . .] experienced by listening'.[66] While a significant body of work has focused on Woolf's overall interest in sound and in 'auscultation', I'd like to link these observations here – the socio-political elements of sound, its ability to intrude upon the individual, and Woolf's fidelity to rendering 'lived' sound – to the anticipatory context of the novel's simultaneously interwar and pre-war setting: to the contemporaneous enskillment of urban listening and the way the expectation of air war trained civilians to locate space through sound.[67]

This point is clearest when we recall that one of Woolf's strongest memories of the First World War was of her spending long hours in the darkness of cellars, listening to the Gotha raids above.[68] The fact that she drew on her own wartime diaries to reconstruct the '1917' section of *The Years* – a much re-written chapter that is itself about an air raid, heard from a cellar – gestures toward a way in which wartime audition seems to splinter temporal boundaries between past, present, and future. In that scene, a 'long-drawn hollow sound wailed out,' interrupting the Pargiters' meal and sending them underground. Listening to the guns above, family friend Nicholas traces the location of the violence in the air, noting its migration from Hampstead to the Embankment and beyond. Here, the ear and not the eye experiences the battle in the sky, but the sounds are visualized as aerial space, as environmental: 'Another gun boomed. A sigh of air rushed up with it,' Woolf writes. 'Then there was a violent crack of sound, like the split of lightning in the sky.'[69] In Woolf's novel, war is not only confused with weather; it becomes a part of it, splicing through the atmosphere to render what Peter Sloterdijk would call the 'climatology' of combat wherein violence is waged by, through, over, and with significant impact on the air.[70]

The air raid scene in *The Years* uses the civilian act of listening in, to, and for wartime to peel back the narrative temporality, such that wartime sound gestures towards the past (Woolf's memory of the First World War), the future (of the Second World War, which motivated the writing of *The Years*), and the present (of mobilization and rehearsal for war's eventuality) simultaneously. Rather than referring only to the air raid in question, the scene shows how, in Sykes's argument about wartime acoustemology, sound and listening can structure wartime experience at multiple levels of duration. He identifies three levels in particular: first, the coming together of extant everyday sounds and the new sounds of war can create a sense of temporal displacement; second, particular sounds can become reference points for short durations of violence itself; and third, certain sounds can represent the experience of living through and enduring the entire war itself.[71] In *The Years* the air-raid siren and the Pargiter family's spatialized listening from below seem to indicate all three. The sounds of the air raid amplify the everyday sounds of 'feet tapping on the pavement' and 'shuffling on the stairs', as well as the silence of a shaking spider web, which represents a fragile world on the brink of destruction.[72] This suspended, plural quality of time is underscored by the characters' anxious listening, which has become metonymic of the civilian experience of aerial bombardment itself, where one is haunted by both a past violence and a violence expected to come again. We are very far from Woolf's earlier celebration of urban life in her essay 'Street Haunting' (1927), where the city is an almost wholly visual spectacle, and where the narrative voice is described as 'a central oyster of perceptiveness, an enormous eye'.[73] Here, the autocratic vision, the enormous eye, is undermined, but we are also left with a narrative more akin to that of 'one large frightened ear', as the earlier diarist had put it.

The Pargiter family is thoroughly arrested in this scene, and memories and thoughts about the air raid resurface in the novel thereafter. However, although an air raid 'interrupts cool and consecutive thinking about peace', Woolf does instil a fleeting but critical moment about what peace, not after but *within* war, could look like. When the Pargiters re-emerge upstairs after the guns pass, Eleanor experiences what she hesitatingly calls 'immunity'. The immunity in part derives from her sense of having survived death but also from her having done so with others both like and unlike herself. For while Nicholas has been locating the violence above, Eleanor has been locating his nationality and sexuality. And when she learns that he is Polish, and then homosexual,

> For a second a sharp shiver of repugnance passed over Eleanor's skin as if a knife had sliced it. Then she realized that it touched nothing of importance. The sharp shiver passed. Underneath was – what? [. . .] All the evening, off and on, she had been feeling about him; this, that, and the other; but now all the feelings came together and made one feeling, one whole – liking.[74]

The loss of physical civilian immunity during an air raid has given way to another, more inter-social kind of immunity that would buttress against the ethnic and sexual prejudices driving the war to come. Woolf gives only the briefest of spaces to this insight, making it barely perceptible, like the details of Oswald Mosley's fascist insignia that later flash momentarily in the last chapter of the novel. When North returns from Africa, he notes offhandedly on a London street that 'somebody had chalked a circle on the wall with a jagged line in it'.[75] Thomas Davis has linked Woolf's de-hierarchization of significant details such as these to her anti-didacticism, arguing that her writing

> dramatiz[es] the signs of history written on the surfaces of everyday life [which] require a certain interpretive practice to acquire legibility. In this way, the novel seems to want to train its contemporary readers to look, observe, and read everything with the same intensity before the catastrophe of an historical event.[76]

Such a reading also requires the auditory dimension, for Woolf is training the reader to hear attentively as well; and if to hear is not only to listen but to attend to different perspectives and points of view, then the aurality of *The Years* registers not only psychological preparation within war's inevitability but also the need for empathy and ethical mobilization. The cultivation of nerves, here, is not just about staying calm in the midst of emergency: it is about 'hold[ing] up against urgency so as to think it', to think of what's at stake in urgency and in resisting it.

The Years does not make this an easy or straightforward endeavour. Not only does the narrative obliquity and anti-didacticism make it difficult to understand the extent of Woolf's characters' – or the novel's – intentions; alongside scenes such as the foregoing, others ask us to continually question the novel's own relationship to empathy. In a controversial episode from the 'Present Day' section, Sara sits in her flat, discussing the poetry of Andrew Marvell with North. Their privacy is spoiled by the sounds of a Jewish neighbour taking a bath in the communal bathtub, and both are perplexed and perplexingly disgusted. '"Damn the Jew!" he [North] exclaimed. The thought of a line of grease from a strange man's body on the bath next door disgusted him.'[77] Even at the most basic level, the scene suggests biological as well as literary-cultural contamination, and this is effected within the novel's persistent concern with sonic intrusion.

But this, the most problematic moment in *The Years*, might also be its most instructive. Moments such as these, brief as they are, nevertheless jar: we hear, but we are not acclimatized. While we can certainly read the scene biographically in terms of the debate surrounding Woolf's own thoughts on Jewishness and anti-semitism in her diaries and letters, we can also read it interpretively and consider the possible emotional regimes which govern Sara's and North's reactions, what they can feel and what they can articulate in the casual anti-semitism of the interwar years.[78] The fact that the novel does not spell out for us how we 'should' read, the fact that the novel demands a critical empathy of itself and even of Woolf – we are critical of her empathy even while we may critically empathize with her – is itself an exercise in interpretive practice as a literary air-raid siren sounding against the state of emergency.

In its two-tone warbling between the aerial and the grounded, *The Years* alerts us to the fact that 'the zoom of a hornet' is not just that of rockets or aeroplanes but what Woolf calls in 'Thoughts on Peace' the 'mind-hornet' of authoritarian thinking that curtails inclusivity and acceptance.[79] Such thinking happens not only abroad but at home; it is broadcasted on the loudspeaker, but it whispers in the subconsciousness. To hear this is to hear the self-obscuring logic that drives states of urgency. To hear this is also to hear the primacy of the individual in thinking through whatever representations of war and its biopolitics are presented. Comprised of telling moments which concern particular uses of sound – whether the air raid's zoom of a hornet or the tapping of an intrusive neighbour – *The Years* underscores sound's quality of being necessarily both private and public, both phenomenological and political. Angela Frattarola has argued that the writer's meticulous description of sounds in her early Second World War diaries is possibly an attempt to capture and control the sounds of the darkening and threatening reality of war: 'pop-pop-pop' go the guns, Woolf writes, alongside the 'drone' of planes flying below, and the way they 'hum and saw and buzz all around us'.[80] Although these sounds populate *Between the Acts* most explicitly, *The Years* shows just how much longer this attempt at sonic capture and mobilization had been in the making.

Notes

1 Mass Observation Archives (MOA), 'Memo on the Senses', TC 'Air Raids, 1938–45', TC 23/6/H. Reproduced with permission of Curtis Brown, London on behalf of the Trustees of the Mass Observation Archive © The Trustees of the Mass Observation Archive.
2 MOA, 'Memo on the Senses'.
3 Virginia Woolf, 'Thoughts on Peace in an Air Raid', in *The Essays of Virginia Woolf*, 6 vols, ed. Stuart N. Clarke (London: The Hogarth Press, 2011), VI, pp. 242–8 (242).
4 Geoffrey Bennington, *Interrupting Derrida* (London: Routledge, 2000), p. 165 (my emphasis).
5 See, for example, J. Martin Daughtry, 'Thanatosonics: Ontologies of Acoustic Violence', *Social Text*, 32.3 (2014), 25–51, which analyses the 'ontologies' of wartime sound primarily in terms of moments of violence: acoustic violence 'besieges the sensorium, introducing empathic and corporeal pain into bodies and forcing subjects to confront their radical finitude (p. 25).
6 Jim Sykes, 'Ontologies of Acoustic Endurance: Rethinking Wartime Sound and Listening', *Sound Studies*, 4.1 (2018), 35–60 (p. 37, emphasis original).
7 Ibid., p. 39. For Sykes, who looks in depth at the decades-long Sri Lankan Civil War, wartime audition is not a universal experience, but one ineluctably 'shaped by one's ethnicity, religion, gender and culturally-construed ontologies of sound, listening, personhood, [and] death' (p. 39).
8 Ibid., p. 38.

9 Steven Connor, 'The Modern Auditory "I"', in *Rewriting the Self: Histories from the Renaissance to the Present*, ed. Roy Porter (London: Routledge, 1997), pp. 203–23 (210–11).
10 I expand on the various modulations of temporal fear, which I call chronophobia, from the pre-war to the post-war period, in *British Literature and Culture in Second World Wartime: For the Duration* (Oxford: Oxford University Press, 2020).
11 Lewis Mumford, *The Culture of Cities* (New York: Harcourt Brace, 1938), p. 275.
12 Paul K. Saint-Amour, *Tense Future: Modernism, Total War, Encyclopedic Form* (Oxford: Oxford University Press, 2015), p. 7.
13 Ibid.
14 Mumford, *The Culture of Cities*, p. 275.
15 Quoted in Peter Adey, 'Still Life: The Private Life of an Air Raid', *M/C Journal: A Journal of Media and Culture*, 12.2 (2009). http://journal.media-culture.org.au/index.php/mcjournal/article/view/112 [accessed 8 July 2020].
16 Ibid.
17 Edward Glover, 'Notes on the Psychological Effects of War Conditions on the Civilian Population', *The International Journal of Psychoanalysis*, 23 (1942), 17–36 (p. 19).
18 P. E. Vernon, 'Psychological Effects of Air-Raids', *Journal of Abnormal and Social Psychology*, 36.4 (October 1941), 457–76 (460).
19 Sylvia Townsend Warner, *The Diaries of Sylvia Townsend Warner* (London: Chatto & Windus, 1994), p. 107.
20 Quoted in Amy Bell, 'Landscapes of Fear: Wartime London 1939–1945', *Journal of British Studies*, 84.1 (January 2009), 153–75 (p. 164).
21 For more on the war as expressed through atmospheric and climatological metaphors, see Maud Ellmann, 'Death in the Air: Virginia Woolf and Sylvia Townsend Warner in World War II', in *Virginia Woolf: Writing the World*, ed. Pamela Caughie and Diana Swanson (Clemson: Clemson University Press, 2015), pp. 77–90.
22 Mary Favret, 'Everyday War', *ELH*, 72.3 (2005), 605–33 (p. 608). Favret expands on this argument in *War at a Distance: Romanticism and the Making of Modern Wartime* (Princeton: Princeton University Press, 2009).
23 Favret, 'Everyday War', p. 608.
24 Tom Harrisson, *Living Through the Blitz* (London: William Collins, 1976), p. 39.
25 Glover, 'Notes', pp. 132–46.
26 James G. Mansell, *The Age of Noise in Britain: Hearing Modernity* (Urbana: University of Illinois Press, 2016), p. 170.
27 Ibid.
28 John Langdon-Davies, *Nerves Versus Nazis* (London: John Routledge & Sons, 1940), p. 7.
29 Ibid., p. 29.
30 Ibid., p. 30.
31 Mansell, *The Age of Noise*, p. 171.
32 Bell, 'Landscapes of Fear', p. 155. See William Reddy, *The Navigation of Feeling: A Framework for the History of Emotions* (Cambridge: Cambridge University Press, 2001).

33 Michel Foucault, *Society Must Be Defended*, trans. David Macey (London: Penguin, 2003), p. 245.
34 See the last section of Michel Foucault, *The History of Sexuality, Volume 1: An Introduction*, trans. Robert Hurley (New York: Pantheon Books, 1978), pp. 133–59.
35 Mumford, *The Culture of Cities*, p. 275.
36 Steve Goodman, *Sonic Warfare: Sound, Affect, and the Ecology of Fear* (Cambridge: MIT Press, 2010), p. 66.
37 Ibid.
38 W. McNaught, A. Langmead Robinson and E. H. Walker, 'Air-Raid Sirens', *Musical Times*, 81.1173 (1940), 457–8 (457).
39 Ibid.
40 Ryan Bishop and John Phillips, 'Manufacturing Emergencies', *Theory, Culture & Society*, 19.4 (2002), 91–102 (97).
41 Quoted in Harrisson, *Living Through the Blitz*, p. 95.
42 Graham Greene, *Twenty-One Stories* (London: Penguin, 1983 [1954]), p. 69.
43 Ibid., p. 72.
44 Quoted in Bell, 'Landscapes of Fear', p. 160.
45 'The Girl with the Golden Temper' [advertisement for Kruschen Salts: 'Go back to Kruschen for the duration'], *Daily Mirror* (9 September 1940), 6.
46 Margery Allingham, *The Traitor's Purse* (New York: Bantam Books, 1983 [1940]), p. 5.
47 Graham Greene, *The Ministry of Fear* (London: Vintage, 2003 [1943]), pp. 70–1.
48 Maurice Wright, 'Psychological Emergencies in War Time', *British Medical Journal*, 2.4105 (1939), 576–8.
49 Each chapter takes place in the year of its title. They are: 1880, 1891, 1907, 1908, 1910, 1911, 1913, 1914, 1917, 1918, and the 'Present Day'.
50 For its complex, six-year composition history, see Anna Snaith's introduction to *The Years*, published as part of the Cambridge Edition of the Works of Virginia Woolf (Cambridge: Cambridge University Press, 2012), particularly pp. xxxix–lxxxvii; and Virginia Woolf, *The Pargiters: The Essay-Novel Portion of the Years*, ed. Mitchell A. Leaska (New York: Harcourt Brace Jovanovich, 1978).
51 Virginia Woolf, *The Years*, ed. Jeri Johnson (London: Penguin, 1998), p. 119.
52 Ellmann, 'Death in the Air', p. 79.
53 Woolf, 'The Leaning Tower', in *The Essays of Virginia Woolf*, VI, pp. 259–83 (272).
54 The loudspeaker is also critiqued in 'Thoughts on Peace in an Air Raid', in which she criticizes Britain's own loudspeakers for its own propaganda, decrying Hitler's tyranny while the country itself harbours a 'subconscious Hitlerism' that voices patriarchy (*The Essays*, VI, pp. 242–8 (243)). For more on the use and political culture of the loudspeaker by National Socialism, see Cornelia Epping-Jäger, 'Hitler's Voice: The Loudspeaker Under National Socialism', *Intermédialités*, 17 (2011), 83–104.
55 Elizabeth F. Evans, 'Air War, Propaganda, and Woolf's Anti-Tyranny Aesthetic', *Modern Fiction Studies*, 59.1 (2013), 53–82.
56 Goodman, *Sonic Warfare*, p. 206n.6.
57 Woolf, *The Years*, ed. Johnson, p. 68.
58 Ibid., p. 81.

59 Virginia Woolf, *The Diary of Virginia Woolf*, 5 vols, ed. Anne Olivier Bell and Andrew McNeillie (London: The Hogarth Press, 1984), V, p. 148. For a concise overview of the intertwined conception of *The Years* and *Three Guineas*, see Anna Snaith, 'Late Virginia Woolf', *Oxford Handbooks Online* (2015). 10.1093/oxfordhb/9780199935338.013.28 [accessed 28 July 2020].
60 Woolf, *The Years*, p. 164.
61 Ibid., p. 166.
62 Ibid., pp. 167, 169, 172.
63 Ibid., p. 170.
64 Ibid., p. 175.
65 Snaith, 'Late'.
66 Snaith, Introduction, in *The Years*, p. xliv; Elicia Clements, 'Reconfigured Terrain: Aural Architecture in *Jacob's Room* and *The Years*', in *Woolf and the City: Selected Papers from the Nineteenth Annual Conference on Virginia Woolf*, ed. Elizabeth F. Evans and Sarah E. Cornish (Clemson: Clemson University Press, 2010), pp. 71–6 (71).
67 Melba Cuddy-Keane uses the term 'auscultation, the action of listening' as a narratological term akin to focalization 'for analysing the reception of sound in literary texts' (Melba Cuddy-Keane, 'Virginia Woolf, Sound Technologies, and the New Aurality', in *Virginia Woolf in the Age of Mechanical Reproduction*, ed. Pamela L. Caughie (London: Garland, 2000), pp. 69–96 (71)).
68 Hermione Lee, *Virginia Woolf* (London: Vintage, 1997), p. 346.
69 Woolf, *The Years*, p. 212.
70 Peter Sloterdijk, *The Terror of the Air* (Cambridge: MIT Press, 2009).
71 Sykes, 'Ontologies of Acoustic Endurance', p. 40.
72 Woolf, *The Years*, pp. 211, 212.
73 Woolf, 'Street Haunting: A London Adventure', in *The Essays of Virginia Woolf*, 6 vols, ed. Andrew McNeillie (London: The Hogarth Press, 1994), IV, pp. 480–91 (481).
74 Woolf, *The Years*, pp. 217–18.
75 Ibid., p. 227.
76 Thomas Davis, 'The Historical Novel at History's End: Virginia Woolf's *The Years*', *Twentieth Century Literature*, 60.1 (2014), 1–26 (18).
77 Woolf, *The Years*, p. 249.
78 Maren Linett points out that Woolf may have covertly blamed Leonard, who was Jewish, for the way her marriage to him intensified and multiplied the threat that tyranny posed to her own safety: that is, after Hitler's rise to power, she herself would be in danger because of her relationship with Leonard. In the same way that the sounds of a Jew intrude upon Sara and North's private conferences, Linett argues, Woolf seems to be obliquely acknowledging and working through a similar intrusion of politics on private life (Maren Linett, 'The Jew in the Bath: Imperilled Imagination in Woolf's *The Years*', *Modern Fiction Studies*, 48.2 (2002), 341–61 (p. 356)). For more on early twentieth-century anti-semitism in Britain, see Lara Trubowitz, *Civil Antisemitism, Modernism, and British Culture, 1902–1939* (Basingstoke: Palgrave Macmillan, 2012).
79 Woolf, 'Thoughts on Peace in an Air Raid', p. 244.
80 Angela Frattarola, 'Listening for "Found Sound" Samples in the Novels of Virginia Woolf', *Woolf Studies Annual*, 11 (2005), 133–59 (p. 156); Woolf, *Diary*, V, pp. 285, 313, 311.

6 Sleuthing Deafness in Detective Fiction

Edward Allen

The thirtieth instalment in Colin Dexter's award-winning TV series, *Inspector Morse*, commences in B minor.[1]

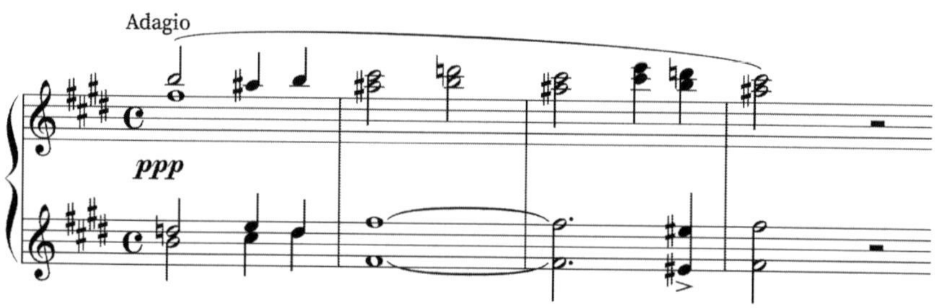

Figure 6.1 Opening scene, *The Daughters of Cain* (1996)

Two women convene in a hospital car park, one smoking, the other chivvying. The nature of their relationship is hard to verify. Then we cut across Oxford to a procession of dons, gowned and chortling, as oblivious as the women were to the incidental music, which has modulated now and softened jauntily into its stride.

Figure 6.2 Opening scene continued, *The Daughters of Cain*

DOI: 10.4324/9780429291616-8

Only then, with another cut, is Morse introduced; and only then – as Morse forsakes his Scotch for the remote control – do we realise that the source of the music is diegetic. As intros go, the prelude to Verdi's *La Traviata* strikes an ambiguous chord, even as the strings would appear – languidly at first, then with a dainty turn – to harmonise the spires and quads, mahogany and candelabra that constitute the opening credits. Unlike the inspector's signature theme, the Verdi does not quite spell things out, but if you're familiar with the opera, as Morse is, you'll guess the plot's about to thicken.[2] Untimely deaths make this storyworld tick, for this is an Oxford in which opera comes true, an Oxford in which crimes of passion and fits of vengeance arise as regularly as essay deadlines.[3] A terminal illness, gambling, and the arrival of a sex worker fulfil to some degree the expectations established by the snippet of *La Traviata*, though it must be said that Morse and Lewis soon have plenty more to contend with, including an antiquated knife wound and a set of bloodied footprints (size 10). We viewers have our work cut out for us, too, a type of meta-detective work that should be differentiated, but is not different in kind, from the efforts of Thames Valley Police. This is the type of work, conducted behind the scenes, that involves evaluating facts, incidents, and clues to which the police will only later, or may never, be made privy. Getting ahead of the game, narratively speaking, would be one way to put it.

One such instance of meta-detective work in *The Daughters of Cain* pertains to a hospital appointment. It is a short and unremarkable exchange – 12½ seconds in length, 51 minutes into the episode – and hardly the kind of thing to interest or detain Morse and Lewis. (They, in any case, are busy elsewhere, last seen motoring up the Cowley Road.) A nurse rises from behind a desk and addresses the motley crowd in outpatients: 'Mr Brooks? Mr Edward Brooks? . . . Mr Brooks? . . . ' Time and NHS waiting lists wait for no man, so the nurse moves down the roster to a Mr Humphries, with rather more success, and there the scene concludes. The point of this scene, or non-scene, is precisely to show the no-show of Mr Brooks. He is high on the police's list, too – for murder as well as for domestic battery and drug dealing – and his failure to make the early-morning appointment may be the first clue that someone has taken the law into her own hands. But as well as prompting us to suspect just for a moment that we have stolen a march on Morse and Lewis, who must wait a few hours more to be shown their suspect bundled up in a roll of carpet, the nurse's unanswered appeals may have a distracting effect, too. One person in the waiting room *does* react visibly to the sound of 'Mr Brooks?', and that person is the person who turns out to be Mr Humphries. This in itself isn't a distraction; on the contrary, the sight of a man looking round to see what all the fuss is about (*Mr Brooks? Who is he?* ***Where*** *is he? I'd happily take his place!*) is intended to render the absence conspicuous. No, the distracting aspect of this bit of stage business has to do with the dual identity of this not-so innocent bystander: *Mr Humphries is Colin Dexter.*

Mystery solved? Not quite.

Seasoned viewers will know that the creator of Morse liked a cameo. In the manner of Alfred Hitchcock, Dexter usually found a way to leave his mark as an extra – mingling, drinking, reading, watching – and usually in close proximity to Morse. His cameo in *The Daughters of Cain* is more complicated. With his back to the nurse who's calling, first for Brooks and then for him (Mr Humphries), Dexter's bit part depends on his tuning into the frequency of a sound without being able to set eyes on its source. This is something most of us do instinctively, but not Colin Dexter. Colin Dexter was deaf – profoundly deaf, had been since his mid-twenties – so his turn as Mr Humphries is something of a puzzle.[4] At the time of filming, digital signal processing (DSP) had not yet supplanted the analogue linear hearing aid, though it is just possible that a mid-nineties device – maybe a 'hybrid' one – enabled Dexter to navigate this acoustically-complex scene.[5] More likely, I think, is the assistance of someone from behind the camera, cueing Dexter in to facilitate his involvement in the bumpy sequence of call and response.

In any case, Dexter's seeming to 'pass' as a hearing person is more significant than a viewing in real-time of this scene or of the episode *in toto* is likely to convey.[6] Because the thing *The Daughters of Cain* should leave you in no doubt about is that deaf people matter. Early in the published text, a key witness very nearly goes unheeded on account of her deafness: although Morse reckons she is a '"phoney from A to Z"', Laura Wynne-Wilson's line of attack – '"Deaf people do *not* require excessive volume – they require only clarity of speech and appropriate lip-movement"' – is spot on.[7] No less important to the plot is KC ('Kevin Costyn; Karate Champion; King of the Condoms').[8] His partial deafness remains a secret, true enough, but it's the fact of his having to sit at the front of class that endears him to his teacher, Julia Stevens, and she it is who persuades him to become an accessory to the murder of Edward Brooks. Dexter's, you might say, is a storyworld in which deafness counts even when it doesn't. Even when it's not central to the plotting, that's to say, you know it is there still, subtly informing the ways people think and act: a source of annoyance for those who'd rather their underlings hadn't noticed ('"Get a hearing aid, sir"'), and the perfect ploy for those who feign it because they believe it's the best way to go *un*noticed.[9] But not in the TV adaptation of *The Daughters of Cain*. Gone are the curtain-twitching antics of Laura Wynne-Wilson; gone is the reference to KC's hidden disability; and there, 51 minutes into the episode, is wonderful Colin Dexter looking for all the world as though he alone can hear the muffled death throes of Mr Brooks. Why?

This is not a question to which I can provide a conclusive answer. Stories are tailored and repackaged for TV; edits happen; things get lost. But what an unfortunate thing to lose in this case. My impression is that small-screen detective dramas have done a patchy job of representing certain types of disability, with the result that many of us have grown up with a warped sense of the category's parameters. Any attempt to survey the genre would have to notice a trend of characters for whom mobility is a visible and chronic problem (Baptiste, House, Ironside, Strike).[10] Perhaps then you would turn to those, less visibly but just as vitally, in whom a growing sensitivity to the nature of mental health and neurodiversity has found embodiment (Marcella, Norén, Poirot, Sherlock).[11] At the same time, it will be clear that television has done much to exaggerate a truth that was there but not always articulated as such in Golden Age detective fiction: to get on as a sleuth, you've got to be something else. Outcast, underdog, interloper, recluse. Sometimes this is realised rather literally on TV – think of the hermit-like dwellings in *Cadfael* (an abbey cell), *Hinterland* (a caravan), and *Jonathan Creek* (a windmill) – but more often than not, a detective's singularity cannot be placed in such concrete terms. He may be a man of colour at Serious Crime (*Luther*) or an aristocrat at Scotland Yard (*The Inspector Lynley Mysteries*); she may be a Brit who talks like a North American (*Killing Eve*) or a woman supposedly doing a man's job (*Prime Suspect*). If difference makes for good drama, it also makes a good detective.

But isn't this to commit a category error? Some would maintain that disability and cultural difference are different kinds of things, and that to confuse or elide the two risks 'missing the meaning of disabilities'; such meaning may be tricky to pinpoint, so this 'medical' argument goes, but its proper, practical interpretation must take account of the fact that 'disability encompasses extreme heterogeneity of type and severity'.[12] Others, by contrast, following the 'social model', have made the case for separating disability from impairment and thus for theorising the ways individuals are disabled, not by their bodies but by the society that seeks to exclude or marginalise them through so many structures, attitudes, and processes.[13] It follows that disabled people – rather than people with disabilities – should be thought to constitute a minority category, firmly situated in the discourse of identity politics.[14] There is not space in the current chapter to do this debate justice; others have done so, and continue to, as my footnotes make plain. What I *can* do, prompted by Colin Dexter, is appraise the value of a third way outlined in the introduction to this volume. What would a 'cultural' approach to detective fiction and disability look like? The clues come thick and fast in Susannah B. Mintz's recent monograph, *The Disabled Detective* (2020). No critic to date has made such a compelling case for imaginative literature's contribution to critical disability studies; she does so by averring something like a synergy between the structural problems with which disability scholars find themselves engaged and the sorts of ethico-formal questions raised by a genre so explicitly invested in matters of law and order. In keeping with my own impression, Mintz also recognises that TV has played a significant part in determining the course of this conversation. That 'we watch more than we read' today is not a reason to grieve, but one imagines (and Mintz implies) that our viewing habits must have something to tell us about the changing status of 'sensory' impairment in the popular imagination.[15] How could they not when the process of intermedial transposition – novel to Netflix episode, print to audio-visual – is bound to loosen the very nuts and bolts that allow detective fiction to function? And by function, of course, I mean *mal*function, too: to suspect that the senses are in jeopardy, at risk of compromise or exposure, misuse or re-invention – and that they need to be, moreover, for things to happen – is to apprehend the real drama of detective fiction.

Without wishing to recycle Mintz's readings, I do intend to cover some of the same conceptual ground in this chapter. And though differing in emphasis, my thinking could also be characterised in terms of 'spatial choreography'; like Mintz, I want to know how knowledge is 'produced by differing interactions in and with' the spaces revealed through detective fiction.[16] A certain way of knowing comes with the territory, you might say. But there's something else I want to know, and that has to do with a reversal of the scenario Mintz has in mind. What happens when disability is embodied by someone other than a detective? Are the effects of impairment in a witness, crook, or victim destined to generate different kinds of knowledge to which a disabled sleuth has access? What, if so, is the currency of such knowledge? To pursue these questions is to test one of the central tenets of the cultural model, which comes down to constructing a theory of epistemology, and to identifying its praxis, in intersubjective terms. As Anne Waldschmidt puts it, 'the cultural model of dis/ability implies that the individual and collective subjectivities of "disabled" and "non-disabled" persons are interdependent'; it stands to reason, she continues, that research will not only attend therefore to the lived experiences of disabled people but also 'compare them with those of persons considered non-disabled'.[17]

For the purposes of this chapter, I take Waldschmidt's proposal as a useful provocation and not, I hasten to add, as an encouragement to dispense with *The Disabled Detective*. It would be rash indeed to bypass Drury Lane, that neglected but most prolix of deaf detectives, whose talent for 'the theatrical reveal' in the 1930s may even have surpassed that of Hercule Poirot, as Mintz suggests.[18] But Lane, I think, should provide a point of departure rather than a destination for this chapter, as for any further work on this subject. As we'll see in the next section, there are precedents for the disabled criminologist, and particularly for the kind who makes a show of his mystic talents ('"I'm scarcely a seer, Inspector,"'

Lane is heard to advise in one novel, tellingly 'with a smile').[19] No less compelling than such precedents, I want to suggest, are the hearing detectives who come to rely on disabled actors. This is what it means to take Waldschmidt's 'cultural' proposal as a prompt, not to de-centre disability but to see it materialising around the crime scene in different forms, sometimes in ways that would appear to arrest, frustrate, or counterpoint the dance of the sleuth's intellect.

Enabling Genius

It is customary in studies of this sort to recall C. Auguste Dupin and Sherlock Holmes, the great minds of classical sleuthing. Taken together, their careers have come to represent the origin story of detective fiction, their methods a model the world over for genius-led ratiocination.[20] You could be forgiven for thinking that no-one had thought to think, or to think hard, before 1841. This itself is a fiction, of course, but one that is hard to resist as you retrace the steps of today's 'defective detective' back into the Golden Age and further still to the murky days of potboilers and penny dreadfuls.[21] Is it foolish to go off-piste? To do so would surely be to doubt the indubitable: any detective worth his salt has a healthy disrespect for authority, this much we know, and it's been this way for as long as there's been a police force to josh with and outsmart. In even their most generous moods, Dupin and Holmes are not slow to declare that most officers lack 'method' and 'imagination'.[22] Yet other ways of telling the genre's history are possible, not just by casting back to classical precedents – to Oedipus or Hamlet, to the mystic or knight-errant – but by effecting a swerve into storyworlds less strictly determined by the game-changing logic of a Dupin.[23] More than disclosing a litter of counterfactual histories, some such ways will throw into relief the things we think we know about the genre, including the ways it gets to know things, to what ends, and from whom.

Long before it was fashionable to deliberate the epistemological framework of genre fiction, readers were being encouraged in terms no less imposing to engage their detective faculties. '[T]here is a demand in almost every page that relates any new act of his, for the finest and closest disquisitions that learning can make upon human nature, to account how those acts could be done by him'. Thus begins one account of 1720 (author unknown) of Duncan Campbell, a deaf Scotsman well known at the time for his ability to conjure names and to foretell love affairs in the manner of a clairvoyant. 'If ever the life of any man under the sun was remarkable,' so this account boasts, 'this Mr. Duncan Campbell's, which I am going to treat upon, is so to a very eminent degree'.[24] One of the enduring interests of *The History* has to do with this gushing style of narration, which does what it can to reel readers in even as it encourages them, in step with London's 'angelic tribe', to circle religiously around the 'heavenly youth'.[25] Part ceremony, part circus, the depiction of Campbell as an object of curiosity and desire has garnered fresh attention among recent critics on account of its intersectional construction of identity: *The History*, in Jason S. Farr's reading, is a work that seeks to establish 'the deaf prophet as a queer, heroic harbinger of a new linguistic order', a 'marvelous persona' from whom 'libertine indulgence' and 'divine wisdom' appear to emanate in more or less equal measure.[26] Farr makes a convincing case in this regard for the 'novel's polarizing tendency', and for the ways a variety of sign-systems – gestural language, fingerspelling, and print – might be thought to have come between the 'second-sighted' prophet and his followers.[27] But what of that generic classification? 'Novel' doesn't feel quite right. This, Farr admits, is not only because the *History*-cum-biography was evidently written for readers well read in travel literature.[28] It is also because, as narrative actors go, the one at the centre of *The History* seems so tantalisingly indefinite.

For Duncan Campbell puzzles. He puzzles, in one sense, because that's what puzzlers do: he is a problem-solver, assisted by a kind of 'genius' that is at once mysterious, palpable, celestial, human, good, enterprising, little, great, and nigh-on impossible to pin down. On the one hand, we are to believe that it is 'by dint of his own genius' – thanks, that is, to some sort of 'inward' talent – that Campbell has taught others to decipher the world as he does, to 'read, write, and converse', with a heightened sensory apparatus.[29] On the other, this 'genius' is understood to be on the move, not intrinsic but fugitive in spirit; it is a kind of 'attending genius', a 'mediate being', which is given to 'whisper[ing]' in Campbell's otherwise unhearing ear as he seeks to make his listeners listen.[30] His is an elusive genius, but it's *il*lusive, too, ever vulnerable to suspicion, and it's for this reason that he might be said to puzzle in another way. Campbell also puzzles in the sense that he remains an 'enigma', both to those who solicit his help – singletons and merchants, hopefuls and no-hopers – and to us. His 'amazing performances are of such a nature', we learn, 'that they can't be well and clearly explained', so much so that his method – 'divination' or 'natural magic'? – is prone to generate 'incredible riddles' far stranger than the mysteries of a client's aching heart.[31]

But for the pleasure it elicits, the type of narrative logic to which I'm pointing – mystery → solution → mystery – would appear to be self-defeating. As it is, this enfolding pattern might also be thought to constitute a positive feedback loop. A genius beset by genius, a puzzler who puzzles and puzzles: Campbell has a 'gift' – the word haunts *The History* – and it's a gift that keeps on giving.[32] It might have done in actual fact, one suspects, had someone in the early 1700s spied a gap in the literary market, for the gift at stake in the case of this deaf 'oracle' – a captivating blend of insight, intuition, ingenuity, and glitz – has serial potential.[33] Poe sensed it in Dupin as he set him to work once more in 'The Mystery of Marie Rogêt' (1842–43); and it's a potential Doyle would make no bones about realising in Holmes (not to begin with, anyway) in the days of the shilling shocker. In a number of ways, but especially in their complex relation to the concept of genius – a relation that is at once an embodiment and a socially determined projection – those nineteenth-century sleuths bear an uncanny resemblance to Duncan Campbell, the 'miracle-monger and impostor' *par excellence*.[34]

This resemblance may heighten our sense that the secret to Dupin's and Holmes's success, to say nothing of Drury Lane's, has a longer history. It may encourage us to speculate that some of the things we take to be essential markers of the latter-day sleuth's personality – the devil-may-care attitude to professional policing, for instance – have been overdetermined, perhaps even to the point that we no longer think of these characters *as* characters. What difference would it make to disband the cult of personality and to approach Holmes once again as a cipher, a puzzle unto himself? Perhaps a great deal. None of this, even so, should be allowed to obscure the possibility that Campbell's gift, and his unlikely gift to detective fiction, can be explained without recourse to his immediate and most obvious successors. A proper gift, as Lewis Hyde observes in his classic essay, effects a 'transformation' in the recipient. '[T]rue teachings' do this, he says – he's thinking of 'the power of example' – and Campbell's ministrations do this, too, to varying degrees; sometimes they may only appear to simulate the spiritual force of the teachings Hyde has in mind, but their capacity to change people, to stir things up – to bring about 'a great suspense of mind' – is readily apparent in *The History*.[35] Yet Hyde would urge us one step further in this thinking. A thing is likely to attain the status of a gift, he argues, as and when it sets things in motion. 'The gift must always move', he says, must be allowed to gain 'momentum', and it does this by prompting the recipient to relinquish something.[36] 'A man may wonder what will come in return for his gift', Hyde admits, 'but he is not supposed to bring it up'; this is what makes 'each donation an act of social faith'.[37] In a world habituated to the politics of bartering and commodity-based transactions, such is the mystery of gift exchange.

And such is the mystery of deafness, I want to argue, in the context of detective fiction. Disability *is* transformative, and it can be enabling, too. The risk of developing this thought, I recognise, is that the talented Mr Campbell may begin to resemble a 'supercrip' in my telling – gifted, indomitable, and finally overstretched – a victim of the 'process by which people with disabilities are portrayed as "noble," "heroic," and "special"'.[38] This risk seems all the more palpable when you consider the popularity today of stories in which disabilities pass for superpowers. Just think of the *X-Men* franchise: it is easy to picture a version of Campbell rocking up at Charles Xavier's 'school for gifted youngsters', nervous at first, perhaps, but finally glad to have found a niche among others similarly disabled by polite society.[39] Yet I do want to pursue this line of thought, to figure out a third way, because it's not clear to me that Campbell belongs to either camp, sub- or superhuman. If the problem with supercrip discourse as a revisionist intervention boils down to a sort of robotic fantasy – a scenario, according to Alice Hall, whereby 'technologised and superhuman power eclipses any sense of human vulnerability' – then Campbell's story encourages us to see precisely that sense coming in and out of focus, blurred sometimes but never wholly occluded by the mobilisation of his deafness.[40] In Hyde's particular understanding of the word, this deafness is a gift: for all his theatrics and ingenuity, Campbell resolves to do 'spiritual good' with it, in the first instance by prompting his interlocutors to unburden themselves; this is a 'reciprocal' act, in Hyde's terms, but hardly a simple one, given the complicated texture of interaction.[41] At the same time, a person bearing gifts, especially a gift such as Campbell's, grows used to being suspected. Where others professing to serve the public might enjoy a 'feeling-bond' – Hyde's term for the sensation of a connection forged through gift exchange – Campbell's capacity for 'penetration and insight' is always subject to scrutiny.[42] He is not what you'd call a trusted source.

Lip Service

My harking back to the example of Duncan Campbell was prompted by an urge to think laterally about the criteria we commonly summon to tell the history of detective fiction. In allowing such criteria to work loose, though not to come apart entirely, we may begin to glimpse an alternative throughline: this is not by way of scoring a direct counterpoint to the sort of story that spotlights '"the irregular pioneer"' – a phrase of Sherlock Holmes's that may or may not do well to describe Campbell – but with a view to pulling focus, the better to appreciate the lengths another sort of story has gone to over the years to inspecting the very modes of perception through which clues are formed as such and according to which evidence is identified and filtered, judged admissible or panned as hearsay.[43] Following in the footsteps of Duncan Campbell is Montgomery Starr, the unlikely hero of Gouverneur Morris's 'The Man Who Played God' (1912). The 'immensely discontented' Starr may at first be more reluctant than Campbell to make anything of his deafness – to adapt it or to adapt to it – and yet he is to prove just as 'gifted' as he reconciles himself to flexing his newly acquired lip-reading skills:

> Starr had discovered that by the aid of his field-glasses, he could, as it were, read the minds of men and women and little children. And that when he wished he could enter into and be a power for good in their lives. [. . .] Through the field-glasses Starr 'heard with his eyes' burglaries planned and seductions. And to the burglarious and to the seducer he sent, by the hand of his valet, words of knowledge and warning that struck them cold with fear. Or he happened upon genuine romances and abetted them, or he interfered between two that quarreled and made their peace for them.[44]

Every private eye needs a sidekick – every wannabe god an emissary – and Starr is no different as he takes to 'eavesdropping' from the vantage of his roof: he would be nowhere without Spring, his trusty valet, who springs forth with 'miracles of distribution' to impart to the mere mortals below.[45] It hardly needs stating that the thing Spring would never do is disabuse Starr of his grandiose delusion; nor does he think it right to apprise his master of Mrs Starr's 'unhallowed leanings' to an old flame by the name of Bobbie Delorme.[46] Having spied an assignation between the two, Starr ends the story – it comes this abruptly – by ending it all.

There are a number of ways one could read 'The Man Who Played God'. It is hard to imagine a more definitive conclusion to a tale of hubris – 'He leaned farther and farther from the parapet of the roof, until suddenly the stone sidewalk rose at him' – a downfall as literal as it is moralistic. The same might be said of an alternative conclusion, of course, for Montgomery Starr's suicide could be construed as an act of self-sacrifice motivated precisely by a too-literal interpretation of the signs.[47] '"He would do *anything* in the world to make me happy"', Mrs Starr confides to Delorme in those last fateful seconds, little knowing the lengths the man who wants to make her happy will actually go, faithful to the last and to the letter.[48] It all hinges on that '"*anything*"'. There's no mistaking Mrs Starr's emphasis; Delorme gets it, a reader gets it. And yet a reader gets something else, I suspect, because the italics can't help but nudge you to see the way words have got to Starr. As if 'hear[ing] with his eyes' is not miraculous enough – the words have got to him, as to any lip-reader, against the odds – Starr ends the tale by demonstrating that the *meaning* has got to him, too. Starr has not only absorbed the visemes, in other words – speech sounds made visible – but he's also tuned in to some of the things an expert in pragmatics would tell us lie *beyond* the basic building blocks of language, including the swirl of implication and relevance in which otherwise nondescript utterances come to pass. '"He would do *anything*"', Mrs Starr asserts. The deaf man feels the force of this – he sees it and sees through it – and he takes her at her word.

Since its *Cosmopolitan* publication in 1912, 'The Man Who Played God' has been transposed many times to stage, screen, and radio. In every case – including the occasion of its becoming a vehicle for Liberace in 1955 ('Fabulously yours in his first starring motion picture!') – the arc of the original story looks more pronounced on account of a common plot development: in 1914's stage play, and in every subsequent version, the hero is no longer just a rich man, gradually altered by a degenerative condition, but a professional musician to whom the sudden onset of deafness makes a material, symbolic difference. A reasonable move for the writers to have made, you might think, perhaps even an improvement of sorts, but one that does little to disguise a further twist, typical of Tinseltown philosophy. If dramatic logic dictates that the victim in all this should be a musician, then it stands to reason – for the purposes of emotional payoff and for the pleasure of a decent showstopper – that the musician's altruism should be rewarded in a timely manner. After so many trials and requests, good deeds and charity gigs, the pianist-hero in *Sincerely Yours* finally finds his hearing loss reversed:

[*Crash*]
Tony [*post-op, lying in a hospital bed*]: What was that?
Nurse: I'm sorry, I dropped . . .
Doctor: . . . she dropped the scissors.
Tony [*quietly amazed*]: I heard it . . . I heard it . . . Thank God!
[*Fade-in concert music and applause*][49]

Call it a miracle – praise be! – chalk it up to the wonders of modern medicine. That's Old Hollywood for you. Yet why brood over such adaptations? The point, of course, is that the protagonist's *own* mode of adaptation has been rendered all but redundant by the clinical plot twist. Where, in the original story, the experience of learning to lip-read is understood to have altered the protagonist for good – eye-opening in two senses – all that hard work is wont in versions like *Sincerely Yours* to seem an interlude or curiosity, a provisional measure, a means to an end in a story whose *actual* end trivialises the story's erstwhile commitment to difference. Lip service to lip-reading, you might say.

To be sure, Liberace's all-singing, all-dancing finale would not have been to everyone's taste in 1955. Consider Miss Maud Silver. She prefers the Book of Common Prayer; for her, a perfect day is the day of rest, replete with 'church bells and evensong, Sunday supper, a quiet hour or two by the fireside, a little talk, a pleasant book, music to be summoned with the turning of a knob'.[50] Patricia Wentworth's crime-busting spinster has often been likened to Agatha Christie's, and it's easy to see why. Like Jane Marple, Silver trusts to a habit of analogical reasoning that could be mistaken for wistfulness, eccentricity, or provincialism: in the heat of investigation, Marple frequently has recourse to the history of her village, dear St Mary Mead, not because she thinks it a paradise but because it seems to her a quintessentially ordinary place, riddled with 'so many interesting human problems'; Silver, for her part, has a quaint treasury of 'elevated maxims' – 'Maudie's Moralities', as Scotland Yard likes to joke – which are drawn in large part from the poetry of a former laureate ('"What an expressive word. It reminds me of Lord Tennyson's charming poem . . ."').[51] Nor is Tennyson the only thing she has at her fingertips. For Silver, like Marple, is an inveterate knitter, and therein lies a method as well as a mask. With privileged access to locations otherwise closed to those who come more conspicuously armed, the yarn bearer is ideally situated to cultivate sources and to do so in such a way as to appear, well, not sharp exactly, but softly needling:

'I thought if I came round straight away without losing any time – '
'There is something that has just occurred – something which troubles you?'
'Yes.'
'And you would like to tell me about it?'
'Yes. I'm going to.'
There was a pause. The room was quiet. Miss Silver knitted.[52]

It would seem an unremarkable bit of writing but for the parataxis and its studied sense of rhythm. 'There was a pause. The room was quiet. Miss Silver knitted.' Silver's is a patterned and patterning intelligence that lives as keenly in the pauses as in the inquiring patter.

If Silver finds in knitting a way to craft her conversations, her twenty-seventh case, *The Listening Eye* (1957), poses a more than usually acute challenge to her method. The person to come 'straight away' to her is Paulina Paine, not, as that phrase might suggest, on page 1, but getting on for halfway through chapter 4. The cause of the delay is innocent enough and has to do with providing Paine's backstory. It all began for her with the London Blitz:

> The last sound which had reached her, or would ever reach her in this world, was the crash of the bomb which had brought her office down over head in 1941. She had been buried in the ruins for twenty-four hours, and when they got her out she was stone-deaf, and had so remained. [. . .] It was odd at first to walk down familiar streets where

> the traffic had roared and see it go by as silently as if it were no more than a picture on the screen at a cinema. But she got used to it, and she got very good at her lip-reading. Practising it continually, it had become second nature. Interesting too. She had read some curious things from lips whose words were pitched for the ear of a friend, a lover, or an enemy.[53]

Sound familiar? The use to which Paine puts her new skills might remind us of Montgomery Starr's charitable occupation, particularly in the extent to which the novice lip-reader – a kind of medium *sans frontières* – can be indiscriminate when it comes to motive and affiliation: friend or foe, no-one's safe. And yet there is a difference between Paine and Starr, or between the circumstances of their surveillance, at any rate. Though she no less than he knows the proverb to be true – all's fair in love and war – public perception of lip-reading had altered significantly, to the point of its becoming polarised, by the mid-1940s. In a continuation of a policy trialled in 1918, military hospitals sought during the Second World War to recruit lip-reading instructors to aid the recovery of deafened serviceman, and so to produce a new generation of 'well-qualified speech readers'.[54] At the same time, government departments on both sides of the Atlantic cast a suspicious eye over the practice for fear of its being weaponised. Historians now tend to agree that the greater purpose of cautionary propaganda – 'Loose lips might sink ships', 'Careless talk costs lives' – was to safeguard civilian morale and not, or not only, to prevent the leaking of sensitive information.[55] Yet the Ministry of Information and its US equivalent were not wrong to imagine the latter possibility or to prompt others to do so. If the brains behind a film like *The Next of Kin* (1942) hoped to instil in audiences a fear of being overheard in public, writers sought in the early years of the Cold War to instil a fear no less paralysing: that is, the fear of being overheard by one who's out of earshot.[56] The fear is palpable in David Karp's *One* (1953), whose lip-reading protagonist, Professor Burden, can't help but pity those he's been directed to surveil. 'They acted', he muses, 'like badly frightened animals'.[57]

To move between such stories and to interpolate their oralist conspiracy theories with things we know now of Cold War tradecraft is to trace the evolution of a very particular sort of paranoia.[58] Paulina Paine comes to know this paranoia in *The Listening Eye* and comes to be known by it. In a series of recursive operations, the novel begins with looking: Paine is visiting a gallery and has come to size up a newly painted portrait of herself called *The Listener*. She looks at it ('it looked very well') and looks at herself looking ('there was that kind of look about the eyes'), only to discover that the look betrays her: 'she thought she had got rid of looking as if she were trying to hear all the things that she would never hear again'.[59] Why the legibility of her deafness should matter remains a mystery – who else would know or care? – and she turns her attention to the other end of the gallery, where another artwork ('[it] suggested the explosion of an atom bomb') has caused a man to idle. This turns out to be your classic 'live drop' – think James Bond, *Skyfall*, London's National Gallery – designed to conceal in plain sight an exchange of words between the man and an equally 'ordinary'-looking accomplice.[60] For a moment, it's as though Paine has wandered into a full-blown spy novel.

But this is no spoof, and Paine turns out to be as much a source of paranoia as its casualty. Thanks to those forensic lip-reading skills, she learns of a double plot (a robbery and a murder) and panics. She panics, of course, because she knows the lethal information she has to impart is not only incomplete (elliptical, lacking in concrete particulars) and not only proleptic (for this information pertains to a sequence of future crimes) but inadmissible: the little she feels she has to give the authorities is thinner even than hearsay, because this is the hearsay of a deaf person. To apprise a policeman of her suspicions, she concludes, 'would be quite hopeless. There had been experiences which resembled it closely enough to assure her of that'.[61] Paine is both right and wrong about this: the police will be slow to believe her, but at least one person won't be; one person, indeed, has already made his move. Upon returning to the scene of the non-crime, she gathers that a gallery attendant knows all about her lip-reading skills (just today, he overheard the artist responsible for *The Listener* explaining the portrait's significance), and like all good gallery attendants Mr Pegler has already woven this trivia into a spiel. Much, it must be said, to the interest of a certain visitor, as Pegler is only too happy to explain:

> '"What!" he said. "You don't mean to say she could be standing over there" – and he points back to the seat what he'd been sitting on – "and that she could tell what you and me was talking about just by looking at us!" "Well, sir," I said, "it's funny you should put it that way, for that's just the way Mr Moray put it – him as painted the picture – when he was talking to the gentleman as bought it. Pointed to that very seat he did and said, 'I give you my word,' he said, 'if she was there and we was here, and you was looking her way, she'd read the words off your lips as fast as you said them.'" You wouldn't credit how interested he was, miss, when I told him that'.[62]

The more Pegler tells Paine of his exchange, the more telling and textured his speech-within-speech-within-speech becomes. Candid, voluble, tickled by coincidence ('"'it's funny you should put it that way'"'), the obliging attendant warms to his theme all over again and, in doing so, gives Paine good cause to worry. She knows to worry, not because Pegler's powers of recall seem especially acute – his rough voicing of Mr Moray, a man we know to speak the Queen's English, suggests otherwise – but because he seems *un*reliable in exactly the way wartime propagandists feared. *Careless talk costs lives*. Pegler's, at any rate, will cost Paine hers.

If a detective novelist was hoping to surprise her readers in 1957, killing off a single innocent bystander probably wasn't the way to do it. Not one but three such deaths give rise to the body count in Christie's *A Murder is Announced* (1950): each one is essential to a tale of identity fraud – each one a consequence of someone being judged to know something they shouldn't – such that the unwitting-witness scenario comes to resemble a category in its own right rather than a mere addendum to the conventional whodunit plot. Wentworth's contribution to what must otherwise seem a well-worn genre hinges not on an innocent's unworldly or unknowing acquisition of knowledge but on the practised competence of Paulina Paine. She is no ordinary bystander. And it's critical to the plotting of *The Listening Eye* that we – and Miss Silver – get to see her performance as a witness so carefully choreographed. Between the news in chapter 3 that Pegler has compromised Paine and a report at the end of chapter 7 that Paine's been involved in a fatal hit-and-run, Wentworth stages a sequence of information-gathering exercises through which Paine is variously refashioned, first, under Silver's care, as a 'percipient' witness – as one who claims to have gleaned what she knows directly through the application of her senses – and then, by the authorities, as a source of anecdotal evidence rendered all the more vicarious by the fact of Silver reporting it over the phone. Though opening herself in this way to the charge of double hearsay, Silver is understood to stand a better chance than Paine of convincing the police that Paine has witnessed something of significance. But it is not to be. The story amounts to a '"pretty fancy sort of thing"', so say the police – a thing from which '"there isn't a clue"' to be gathered – and Silver must simply bide her time to see the pretty fancy sort of thing reported as fact (in *The Times*, as it happens).[63] For the moment, Silver alone sees Paine for what she is and wakes the following night to find 'the problem [. . .] vividly present'.[64] Neither she nor Wentworth is inclined to disambiguate that suggestive phrase, but perhaps by now we have all the proof we need. The problem, to Silver's mind, is that a deaf woman should be considered an '"informant"' – an expert bystander, unbidden and fully embodied – and that she, Silver, is the only one to have realised it.[65]

Morse Code

Paulina Paine's death would seem a particularly cruel instance of collateral damage but for the fact that Silver gets to her before her murderers do. Left to her own devices, Silver would act immediately on her informant's information, and we're led to understand that she would do so because she knows a credible testimony when she hears one. '"I may have been mistaken in a word or words which would alter the whole sense,"' Paine tells Silver, '"and of course only one side of the conversation reached me."'[66] The lip-reader is nothing if not circumspect – swift to qualify her findings, admirably immune to confirmation bias – which renders the story she's told Silver only more compelling, not less so. But that's not the way detective novels work. Even in the case of time-sensitive information – especially in such cases, perhaps – it simply wouldn't do for a private detective to receive a prompt hearing, or a fair one; she has to be doubted before she can be believed. As sure as day follows night, Silver knows the police are obliged – temperamentally and professionally, and for the purposes of narrative design, above all – to think better of accepting a tip-off or a head start. That she prefers to style herself as a 'private enquiry agent' is neither here nor there: Silver belongs to the private sector, which means that her time will come, but only once law enforcement has shown itself to be, in Raymond Chandler's phrase, 'a very imperfect mechanism'.[67]

If novels of this kind have their routine – a time-stalling dance between the powers that be and the private detective – you've got to wonder whether a person in Paine's position would be better served in the subgenre known as the police procedural. Just ask Nicholas Quinn. His recent appointment to the Foreign Examinations Syndicate in Oxford sets the scene for Colin Dexter's third novel, and he's the first in quite a line of deaf characters to puzzle Inspector Morse. With no Silver or Marple seeking to bend his ear, Morse has no reason not to do what Morse thinks, no reason not to hear or heed the voice within. Yet it's not always easy to judge in Dexter the lie of certain voices or to pinpoint whence the prompt to rethink this theory or that is coming. Are these the inner workings of a conflicted investigator – 'Why not go with the evidence, Morse, and fling your fancies aside?'; 'Go on, Morse. [. . .] Try again, Morse' – or the probing of an impatient third party?[68] *Someone* knows Morse is on the wrong track. And that same someone – we might call it a free indirect blend of self-examination and narrative oversight – must also be responsible in part for having set a reader up to notice or dismiss, or else to miss entirely, facts that Morse finally comes to believe are clues. It is a twofold manipulation. If we are given to understand some way through the book that Morse has been led astray by a particular suspect or piece of information ('Try again, Morse'), we must recognise that we, too, have been led on in certain ways, not to the extent of thinking we know much more than Morse but to the point of beginning to sense (belatedly) that we were made privy to things earlier in the story in such a way as to result in our not knowing what we know – that we know, in other words, but have been slow to *realise* that we know, several things that turn out to be significant.

The principle of manipulation, so carefully sequenced, is not peculiar to Dexter's fiction. Probably all detective novels involve the dovetailing of epistemological crises, ours and the detective's. Yet *The Silent World of Nicholas Quinn* (1977) offers an especially rich example, because the basis of the manipulation – deafness and what we know of it – is a variegated thing, a construct as well as a lived experience. In a figurative sense, deafness amounts to a body of knowledge to which we and then Morse are introduced – what it means to be deaf in 'social' terms – and which we're all manipulated into thinking we've grasped; in another, more literal sense, however, this is a body of knowledge that begins and ends, inescapably, with a body. What it means to be deaf and what it means to be Quinn are not quite the same thing.

Thames Valley Police will take some time to work this out, come the investigation which commences in chapter 5, while we, by contrast, cannot but see a distinction has to be made between the two as Dexter inducts us into Quinn's 'silent world'. On the one hand, it's thanks to affirmative action that Quinn finds employment at the Syndicate, whose constitution compels it to accept such persons if '"'the disabilities of such persons prove not substantially to interfere with the proper discharge of the duties entrusted to them'"'.[69] On the other, just a chapter later and no more than a month or so into the job, Quinn is worrying already that 'his deafness [is going to] be slightly more of a handicap than he'd sometimes assumed': an office environment is hard to navigate, and fielding questions by phone as to what should or should not appear on a particular exam paper can be fraught.[70] Quinn's colleagues are not slow to notice this, given a series of farcical exchanges ('"Hell of a sight more relevant than Elizabeth the First, eh?'" '"Better than a 'list of metaphors', did you say?"'); yet it would be rash to conclude that Quinn thinks himself a liability or a figure of fun, for he believes – through lip-reading – that 'he must have been blessed with a rare gift'.[71] A help and a hindrance, a source of mirth and a condition of possibility ('it was quite wonderful what one *could* do') – the fact and many varied effects of Quinn's deafness require one to get used to making perceptual adjustments, all of which contribute to the sense that we're being misdirected.[72] Is this interloper being groomed to be an accomplice (why else would that fellow in the prologue have advocated so vociferously for Quinn's appointment?), or is 'the innocent-looking Quinn', as he appears to one green-eyed colleague, to be the unwitting provocateur of some crime of passion, 'with his hearing aid, and his wide and guileless eyes'?[73] It is hard to say where we're heading until the very moment Quinn winds up dead, and even then – here's Dexter's dexterity – it's even harder to divine the motive.

Thus Morse enters the fray. Where we have been encouraged hitherto, even compelled, to recognise the mismatch between an institution's experience of deafness and that of one of its employees, Morse seems disposed at first to read as little into the mismatch as possible. In consequence, we bear witness to him playing catch-up – to his learning things we already know in some detail – as well as to his being told things we know are false. When Dr Bartlett, the Syndicate's permanent secretary, admits to Morse to having been '"very impressed by the way [Quinn] coped with his disability"' – this from the mouth of a man who once grimaced ('as though he had swallowed a cupful of vinegar') to recall the way the Syndicate had been hectored into appointing Quinn – we know he's fibbing, and we know his squiffy interviewer's none the wiser.[74] If only Morse knew he were the hero on whom a novel called *The Silent World of Nicholas Quinn* depends! How quick he'd be to put to one side the mysterious green anorak or to give up on the 'buff-coloured' cinema tickets that seem, by some kind of wish fulfilment, to lead straight back to a viewing of *The Nymphomaniac*.[75] Pull yourself together, Morse!

And yet, if Morse is slower than he should be to cotton on to what the novel is and always has been about, he shows himself to be resourceful and imaginative when it counts. He does so in a way that serves not only to make sense of everything we know but to reveal, crucially, some things we didn't know we knew. In a way we may have come to suspect, the key scene turns out to have been a conference at an Oxford hotel in chapter 3 – a 'jamboree' of sorts – at which Quinn and his colleagues were assigned the task of marketing a new history syllabus to a delegation hailing from an oil state in the Middle East.[76] In retrospect, as on first reading, the scene sounds a vocally complex one, in keeping with the kinds of situation in which Quinn's disability always appears most legible: the strained voices and nervous looks of a gathering, the mishearings, the ringing of the 'squat and menacing' telephones.[77] The scene stands out, however, not because it conforms to the expectations established in the Syndicate's offices but because the lip-reader seems to navigate the vital moment with aplomb. As much for our sake as for Quinn's, Dexter is careful to clarify that members of the delegation speak 'a beautifully precise and fluent brand of English' – 'distinct and [. . .] almost childishly comprehensible' – so it seems reasonable to conclude that the fruits of Quinn's eavesdropping are to be taken down in evidence.[78] '[H]e knew that the words must certainly have been whispered *pianissimo*. He felt his heart beat thicker and faster. He must surely have been mistaken . . .'[79]

The natural conclusion, of course, is that Quinn cannot believe his eyes. The fact of this conversation or any other being whispered is immaterial: for the itinerant lip-reader, it's 'as if [the words] were being shouted through a megaphone'.[80] Without having witnessed the scene, Morse finally comes to this conclusion, too, and perhaps atones for the delay by recognising the '"obvious"':

> 'First of all, I think, the single most important fact in the whole case was Quinn's deafness. You see Quinn was not only hard of hearing; he was very very deaf. But we learned that he was quite exceptionally proficient in the art of lip-reading; and I'm quite sure that because he could lip-read so brilliantly Quinn discovered the staggering fact that one of his colleagues was crooked. You see the real sin against the Holy Ghost for anyone in charge of public examinations is to divulge the contents of question papers beforehand; and Quinn discovered that one of his colleagues was doing precisely that.'[81]

Morse believes he's cracked the case, and we probably believe it too, since it corroborates the scene in chapter 3. Morse is just getting started, however:

> '*But*, Lewis, I failed to take into account a much more obvious and much more important implication of Quinn's being deaf. [. . .] Quinn was a marvel at reading from the lips of others – agreed? He might just as well have had ears, really. But he could only, let's say, *hear* what others were saying when he could *see* them.'[82]

As it turns out, Morse is right on every count: he is right to call Quinn's deafness '"the single most important fact"'; he is right to have investigated the *extent* of Quinn's deafness; and he is right to suspect that Quinn was unfortunate enough – just like Paulina Paine – to uncover a criminal scheme by virtue of lip-reading. He is even right to draw out that final implication, and we know he's right – of course he is! – because those tell-tale italics bespeak a gnomic common-sense that sounds irrefutable: '"*Quinn couldn't hear what he didn't see.*"'[83]

'"*At nine-thirty Mr Ackroyd was already dead.*"'[84] '"[N]*o-one in any other part of the train could possibly have murdered Ratchett*"'.[85] '"*On the night in question Miss Arundell's breath was phosphorescent!*"'[86] Nobody delivers a clincher quite in the manner of Poirot, but Morse comes close in the case of Nicholas Quinn. And perhaps it's this, on second thoughts – together with Morse's plan to gather all the suspects for a 'little party-piece' the next morning – that prompts one to wonder whether something so campily staged, fifty pages to go, can be true. For all the ways in which Morse is right, could this be a false ending? Lewis, his sturdy sidekick, remains perplexed, and Morse himself admits (admits to himself, anyway) to 'one or two slight inconsistencies' in his theory, even as a wrongly accused man, Dr Bartlett, sits wonderingly in custody. His line of reasoning indicates a proper sensitivity to the facts of Quinn's deafness, right enough. But *whose* facts? '"[W]e learned that he was quite exceptionally proficient"', Morse recalls, insensible for the moment to the truth that he only knows what he knows about Quinn's proficiency in '"the art of lip-reading"' either because he's heard it from a witness ('"he was quite remarkable"') or because he's put himself in the other man's shoes ('"Lip-reading's absolutely useless when you can't see the person who's talking . . ."').[87] Even to think of it as an '"art"' indicates a sort of oralist misperception, and it's owing to this misperception – a symptom of supercrip thinking, no doubt – that Morse is prompted to suppose that Quinn '"might just as well have had ears, really"'. No, Morse, not really. Had Quinn had working ears, he'd never have found himself straining to catch all those distant 'whispered' words at the conference. But more than this: had Quinn had working ears within earshot of the thing he is supposed to have detected, he would never have mistaken the name on that careless whisperer's lips.

Only by attending a lip-reading class does Morse come to realise his mistake, and Quinn's. 'Frequently, it appeared, the difficult words – the words that the class were puzzled by – began with a "p", or "b", or "m"; and to a lesser extent with "t", "d", or "n"'.[88] Still Morse fails to see it – to visualise the slip-up itself or to grasp the nature of the problem – but suddenly, one day, it comes to him, like the sputtering memory of a silent film. 'Lip-reading might be a wonderful thing, but even the teacher of the class had made a mistake, hadn't she? When he'd asked her . . .'.[89] Morse thinks back to his post-tutorial questioning ('"Could Quinn hear the bell?"') and to its mishearing ('"Could Quinn hear them all?"'); and so he attends to the possibility – knowing now the hazards of consonants and elision (*thebell/themall*) – that Quinn saw what no-one else could have heard, not because others weren't close enough to hear it but because what he saw was never spoken:

> 'I told you the key to this case lay in the fact that Quinn was deaf. And so it was. But I kept on thinking what a marvel he must have become at lip-reading, and I overlooked the most obvious thing of all: that even the best lip-reader in the world is sometimes going to make a few mistakes; and Quinn did just that. [. . .] if you mouth the words "Bartlett" and "Martin", there's very little difference on the lips. The "B" and the "M" are absolutely identical, and the second part of each of the names gets swallowed up in the mouth somewhere. But that's not all. It was *Doctor* Bartlett, and *Donald* Martin. Just try them again. *Very* little difference to see [. . .]'[90]

A few mistakes. Morse has rarely spoken a truer word. He speaks for Quinn, but he speaks for himself, too, in this moment. And perhaps he speaks for us – or for anyone, at any rate, who mistook the sense of a perceived mistake back in chapter 3. 'He must surely have been mistaken . . .'.[91] The sense of that phrase lands differently now, and the 'must surely' with the slightest of bumps: it's no longer easy to know whether Quinn, the lip-reader, couldn't believe his eyes that fateful day, or shouldn't have. Some mysteries go unsolved.

Listener's Remorse

Although he admits later to having had '"an uneasy feeling that [he] might have got it all wrong"', Morse never acknowledges in explicit terms the origin of his error.[92] For him to do so, he would have to recognise the asymmetry of his method. For too much of his investigation into Quinn's death he relies on hearing sources: sources to whom he listens, that is, and sources who know what they know through hearing, for whom the effects of Quinn's deafness have become a type of institutional knowledge. As it is, Morse's late, confusing exchange at the lip-reading class is, to one way of thinking, the most telling exchange in the whole book, not only because it finally gives him what he needs to know but because it also provides an object lesson in what it means, what it *takes*, to be acculturated to a 'silent world' such as Quinn's. In the space of little more than a page, we witness the true reach and limit of Morse's imagination. That he feels, on watching the class unfold, that he's 'suddenly been struck deaf' hardly amounts to a paradigm shift, but it makes a difference to him and to the spirit in which he absorbs the teacher's memory of Quinn. 'Yes, he really had been very deaf indeed – but one wouldn't have guessed; unless, that is, one had experience of these things'.[93] Poirot would have whispered that sub-ordinate clause, but really, the 'unless' tells you all you need to know.

None of this is to imply that Morse isn't the right man for the job. It hardly needs stating that his being readier to seek out the inside track would have made for a shorter (lesser) novel. But there's a serious reason for staying the course with Morse. Time and again, he finds himself mistaken – 'even Morse – especially Morse! – was sometimes wholly wrong' – and therein lies his importance to us, as well as his charm.[94] To be mistaken is not in itself a crime, far from it; the more significant mistake would be for Morse to proceed either as though he's learned nothing or as though he knows it all. He does learn from his mistakes, whilst knowing there's always more to know, such that we might think of the particular source of complexity at stake in this chapter – deafness – as an index of his evolving aptitude as a detective. If on his first outing it suits him to be mistaken for 'a very deaf person with his hearing-aid switched off', Morse shows himself to be open-minded as he moves from one case to the next.[95] Not broad-minded, mind – this is still Morse we're talking about – but bloody-mindedly *open*-minded: open-minded in the special sense conjured by police chiefs when they speak at press conferences of 'keeping an open mind'. Coming so early in the series, the case of Nicholas Quinn is formative in this respect, for it compels Morse to acknowledge a paradox he'd be well advised to police by in future: that no clear corollary about a person's detective skills can be deduced from a case in which that person might have got it right had he not got it wrong, and that a detective would be a fool to deduce that deaf sources are not to be trusted because one once got it wrong when he might have got it right.

In grasping this paradox, Morse is never tempted to pre-judge the deaf community or to judge a deaf actor on anything but her own terms, which means that he remains alive to what she might be capable of. Sometimes this prompts him to go where others haven't or wouldn't, like Number 19, the home of Mrs Arabella Adams in *Death Is Now My Neighbour* (1996): the detective very nearly botches this key witness, but for other reasons, and not because she's '"stone-deaf without her hearing-aid"'.[96] Sometimes, as logic dictates, Morse is led instead to suspect foul play, as in *The Riddle of the Third Mile* (1983) and *The Jewel That Was Ours* (1991): this is not because he thinks deaf actors more capable than hearing ones of criminal activity, but because he knows a connection of the sort Lewis has not quite spelled out in a character profile ('Little dapper bumshious [*sic*] fellow – slightly deaf – pretty secretive') needs to be verified one way or the other.[97] Morse rarely cracks it first time, but he's surely right in each of these cases to believe you underestimate a deaf person at your peril.

The point of this chapter has been to capture some of the ways detective fiction has had its epistemological claims exposed by the enabling and/or disabling effects of deafness. The 'and/or' is important, I think – in contrast to an 'either/or' – because sometimes it *is* both, and that's as true of stories in which deafness counts among the defining features of a suspect, criminal or victim as it of those in which detectives find themselves at odds with the hearing majority. To lean into this 'and/or' may be to register the force of Ria Cheyne's observation that disability cannot be 'neatly categorised' when it comes to the crime genre, and that we should take note of stories which 'generate productive misfits' by spotlighting or upending our expectations as to whom the 'villainous' and 'vulnerable' labels apply.[98] Within certain parameters. As Cheyne helps us to see, all such 'productive misfits' are generated in order to heighten the 'desired feelings' a reader arrives hoping to experience; they point up, rather than ever threaten to make new, the 'patterns of affective response' we've come to expect of a particular genre.[99] To put it in terms of the stories I've strung together here – not all of them 'crime' stories, but stories in which the 'detective' dynamic is key – deafness serves to make a story more puzzling, its solution more ingenious. Even the detective stories that meddle most productively with our expectations of disability, and *The Silent World of Nicholas Quinn* must surely rank among these, as Cheyne herself discerns – yes, even these must be understood to have recruited disability in large part because the disabled person provides so rich a nexus for all the things the host genre is expected to deliver. This is true to some degree of characters who have come down to us in puzzle-texts, to whom a future essay might attend – 'Madonna' in *Hide and Seek* (1854), Ellie and Albert Morgan in 'The Key' (1941), Jane Lambert in *Morality Play* (1995) – but Nicholas Quinn remains the outstanding example.[100] However much of a misfit he may be, his deafness is a perfect fit for detective fiction proper.

A genre in which disability is not always 'neatly categorised' does not a revolution make. Detective fiction is what it is, after all – formulaic, formally conservative – and it would be absurd to expect to find in it characters as enlightened or reconstructed as presentist opinion today dictates. This is not to say that disability studies and detective fiction of a certain vintage haven't important things to share with one another. If detective fiction hesitates to imagine a social order in which deafness amounts not to a disability but to a valuable, highly contingent means of knowing the world – imagine that! – it is not afraid to register the absence of such an order. Dexter's deaf characters feel the absence keenly, though it is never articulated as such. Rather than something that hasn't yet come to pass, what the characters long for is something that's long gone, or going. It's hearing they miss. You sense it in the bursting volubility of Arabella Adams and Laura Wynne-Wilson; in the phony swagger of Kevin Costyn, and the bluster of Jim Strange. You sense it in the wrongly-accused Cedric Downes, in his fiddling and evasions, and in that expression of relief – '"Back in the land of the living!"' – as his back-up aid squeals into operation.[101]

And you sense it in Morse's last case, more starkly than ever, in something that feels like a voiceover:

> [S]ilence, for those who are deaf, is not merely an absence of noise, but is a wholly *passive* silence, in which the potential vibrancy of active silence can never again be appreciated. Deafness is not the brief pregnant silence on the radio when the listener awaits the Greenwich time-signal; deafness is a radio-set that is defunct, its batteries dead and non-renewable.[102]

You don't have to be a detective to detect the frustration in Dexter's elegiac turn: it resides in the snap of those italics; flashes for a moment in the odd flatness of that terminal metaphor. Yet hidden in plain sight of this 'medical' articulation of disability might be its solution. If the deaf in Dexter know the frustration and sadness of which he speaks, it's because they all belong to a hearing storyworld. In such a world, those who miss hearing are mishearing's tools; they put up with choppy conversations, with lip-reading classes and whistling hearing-aids; they're used to the mixed signals of an audist society that shows scant sign of signing. *What if the deaf in this storyworld just long to be Deaf?* This, it seems to me, is one of Dexter's most longstanding riddles. And if we feel closer, twenty years on, to solving it, it must be thanks to the sometimes wayward efforts of Inspector Morse, not in spite of them.

Notes

1 *The Daughters of Cain* (1996), screenplay by Julian Mitchell, directed by Herbert Wise.

2 Through a syncopated inverted pedal note, Barrington Pheloung's theme turns morse code into music.

3 Morse himself recognises as much in episode 19. Of their latest case he remarks to Lewis: 'Almost operatic . . . Greek tragedy . . . You should try the myths sometime, Lewis. Sex is never simple, though. There's pleasure, and then there's payment . . . retribution . . . ' (*Greeks Bearing Gifts* [1991)]).

4 Dexter talks with some pain about the onset of his deafness and attempts to 'cure' it in a late interview with Rob McGibbon ('The Definite Article – Writer's Cut', 8 August 2014: <www.robmcgibbon.com/inspector-morse-author-colin-dexter-definite-article-writers-cut/> [accessed 26 May 2020]). The choreography of the scene in question replicates exactly the kind of arrangement Dexter claims to have found impossible in the end in a teaching setting.

An endnote about Dexter's deafness prompts a specific note about the terms used in this chapter. To capitalise or not to capitalise? That is still, to some minds, the question to which anyone writing about d/Deafness must attend. I opt throughout this chapter for the lowercase in its nounal and adjectival forms for several reasons connected to the criteria outlined by the EASIER project (www.project-easier.eu): that the characters under scrutiny in this chapter are all understood to have lost their hearing, and are consequently minded to see deafness as an impairment; that all these characters, save for one, rely on a suite of oralist measures (hearing-aids, and lip-reading rather than signing); and that – as the cause or consequence of this – none of these characters can be said to have been integrated into the Deaf community. Herein lies a clue to the conclusion I'll come to in due course.

5 See Harry Levitt, 'A Historical Perspective on Digital Hearing Aids: How Digital Technology Has Changed Modern Hearing Aids', *Trends in Amplification*, 11.1 (2007), 7–24.

6 For more on the application of this term in disability studies, see Noel Patrick O'Connell, '"Passing as Normal": Living and Coping with the Stigma of Deafness', *Qualitative Inquiry*, 22.8 (2016), 651–61.

7 Colin Dexter, *The Daughters of Cain* (London: Macmillan, 1994), pp. 38, 29.

8 Ibid., p. xix.

9 Morse is well aware that his superior, DCS Strange, is 'going deaf' (see Dexter, *The Way Through the Woods* (London: Macmillan, 1992), p. 357; and *Death Is Now My Neighbour* (London: Macmillan, 1996), p. 188)). At the same time, Morse knows 'passing' as a deaf person can be advantageous in his line of work (see *Death Is Now My Neighbour*, p. 54; and *The Jewel That Was Ours* (London: Macmillan, 1991), p. 34).

10 In the BBC's *The Missing* and *Baptiste* (2014–21) Julien Baptiste suffers from a limp sustained in the line of duty; in Fox's *House* (2004–12) the titular head

of diagnostics uses a cane in consequence of chronic pain brought on by an aneurism in his right leg; NBC's *Ironside* (1967–75) features the wheelchair-bound Robert T. Ironside; and the detective in Robert Galbraith's *Cormoran Strike* novels – in the process of being adapted for the BBC – is an amputee war veteran.

11 Dissociative identity disorder, Asperger's, obsessive-compulsive disorder, and autism have been attributed respectively (though not without some debate) to the lead characters in ITV's *Marcella* (2016–21), SVT1/DRT1's *The Bridge* (2011–18), ITV's *Poirot* (1989–2013), and the BBC's *Sherlock* (2010–17).

12 Dimitris Anastasiou and James M. Kauffman, 'Disability as Cultural Difference: Implications for Special Education', *Remedial and Special Education*, 33.3 (2012), 139–49 (p. 144).

13 Tom Shakespeare, 'The Social Model of Disability', in *The Disability Studies Reader*, ed. Lennard J. Davis, 4th edition (New York and London: Routledge, 2013), pp. 214–21 (218).

14 Michael Oliver, 'The Social Model in Action: If I Had a Hammer', in *Implementing the Social Model of Disability: Theory and Research*, ed. C. Barnes and G. Mercer (Leeds: Disability Press, 2004), pp. 18–31.

15 Susannah B. Mintz, *The Disabled Detective: Sleuthing Disability in Contemporary Crime Fiction* (London: Bloomsbury, 2020), pp. 20–1.

16 Ibid., pp. 90, 12.

17 Anne Waldschmidt, 'Disability – Culture – Society: Strengths and Weaknesses of a Cultural Model of Dis/ability', *Alter: European Journal of Disability Research*, 12 (2018), 67–80 (75).

18 Mintz, *The Disabled Detective*, p. 69. For Mintz's extended assessment of Drury Lane, see *The Disabled Detective*, pp. 68–75. As Mintz explains, Ellery Queen – the creator of Drury Lane – was the penname of two writers Frederic Dannay (or Daniel Nathan) and Manfred Bennington Lee (or Emanuel Benjamin Lepofsky).

19 Ellery Queen, *The Tragedy of Y* (New York: Avon, 1932), p. 148.

20 'Genius' is the word one tends to read a lot in these parts. See, for example, Mintz, *The Disabled Detective*, p. 13; and Michael Cohen, 'The Detective as Other: The Detective *versus* the Other', in *Diversity and Detective Fiction*, ed. Kathleen Gregory Klein (Bowling Green, OH: Bowling Green State University Popular Press, 1999), pp. 144–57 (154).

21 I borrow the phrase from an anthology whose introduction duly opens with Dupin and Holmes: *The Defective Detective in the Pulps*, ed. Gary Hoppenstand and Ray B. Browne (Bowling Green, OH: Bowling Green State University Popular Press, 1983), pp. 1–7.

22 Edgar Allan Poe, 'The Murders in the Rue Morgue' (1841), in *Poetry and Tales*, ed. Patrick F. Quinn (New York: Library of America), pp. 397–431; Arthur Conan Doyle, 'Silver Blaze' (1892), in *The Penguin Complete Sherlock Holmes* (London: Penguin, 2009), pp. 335–50 (339).

23 The genre's historical origins and influences have been much debated. See, for instance: Michael Cook, *Narratives of Enclosure in Detective Fiction: The Locked Room Mystery* (New York: Palgrave Macmillan, 2011), p. 2; Susan Rowland, *The Sleuth and the Goddess: Hestia, Artemis, Athena and Aphrodite*

in Women's Detective Fiction (New York: Routledge, 2020); and *Christianity and the Detective Story*, ed. Anya Morlan and Walter Raubicheck (Newcastle-upon-Tyne: Cambridge Scholars, 2013).

24 [Anon.], *The History of the Life and Adventures of Mr. Duncan Campbell* (London: E. Curll, 1720), p. 2.

25 Ibid., p. 99.

26 Jason S. Farr, *Novel Bodies: Disability and Sexuality in Eighteenth-Century British Literature* (Lewisburg, PA: Bucknell University Press, 2019), pp. 36–70 (55, 45). See also Christopher Krentz, 'Duncan Campbell and the Discourses of Deafness', *Prose Studies*, 27.1–2 (2005), 39–52.

27 [Anon.], *The History*, p. 28.

28 Farr, *Novel Bodies*, pp. 42, 66.

29 [Anon.], *The History*, pp. 4, 99.

30 Ibid., p. 79.

31 Ibid., pp. 157, 175.

32 Ibid., pp. 11, 20, 100, 130, 133, 148.

33 Ibid., p. 48.

34 Ibid., p. 5.

35 Lewis Hyde, 'The Gift Must Always Move', *CoEvolution Quarterly*, 35 (1981), 10–31 (11); [Anon.], *The History*, p. 119. The double-sense in which I'm treating 'gift' – as 'talent' and 'intuition' – is covered by Hyde in his extended account: *The Gift: Creativity and the Artist in the Modern World* (New York: Random House, 2007), p. xvi.

36 Hyde, 'The Gift Must Always Move', pp. 12, 15.

37 Ibid., pp. 18, 19.

38 Lennard J. Davis, *Enforcing Normalcy: Disability, Deafness, and the Body* (London and New York: Verso, 1995), p. 106.

39 The plaque on the gate of Xavier's mansion – a haven for mutants – features regularly in the film series. It is the time-warping outing of 2014, *X-Men: Days of Future Past*, in which I can imagine Campbell finding a home. For more on the intersection of disability and superpowers, see: José Alaniz, *Death, Disability, and the Superhero: The Silver Age and Beyond* (Jackson: University Press of Mississippi, 2014).

40 Alice Hall, *Literature and Disability* (London and New York: Routledge, 2016), p. 11.

41 [Anon.], *The History*, p. 175; Hyde, 'The Gift Must Always Move', p. 19.

42 [Anon.], *The History*, p. 48; Hyde, 'The Gift Must Always Move', p. 11.

43 Doyle, 'The Adventure of the Missing Three-Quarter' (1904), in *The Penguin Complete Sherlock Holmes*, pp. 622–35 (629).

44 Gouverneur Morris, 'The Man Who Played God', *Cosmopolitan Magazine*, 52 (December 1911–May 1912), 278–89 (278, 279, 284–5).

45 Ibid., 283.

46 Ibid., 285, 289.

47 Ibid., 289.

48 Ibid.

49 *Sincerely Yours* (1955), screenplay by Irving Wallace, directed by Gordon Douglas. The first adaptation of Morris's story – and the first to depart from

script by restoring the protagonist's hearing – was Jules Eckert Goodman's play *The Silent Voice* (1914), which was duly adapted for the screen by I. K. Freedman and Eve Unsell (*The Silent Voice* (1915)). This was followed by another silent screen version, *The Man Who Played God* (1922), which retained Goodman's corrective plot twist, and by a remake of that version in 1932 (starring George Arliss) *without* the plot twist. Radio plays followed in 1938 and 1942, before the story was treated in 1955 to a significant makeover (including a surgical operation which leads to the restoration of the protagonist's hearing, as the extract here depicts) when Liberace was cast in *Sincerely Yours*.

50 Patricia Wentworth, *Vanishing Point* (London: Hodder and Stoughton, 2006 [1953]), p. 126. (Patricia Wentworth was the penname of Dora Amy Turnbull.)

51 Agatha Christie, *A Caribbean Mystery* (London: Harper Collins, 2022 [1964]), p. 6; Wentworth, *Anna, Where Are You?* (London: Hodder and Stoughton, 2006 [1951]), p. 334; Wentworth, *The Fingerprint* (London: Hodder and Stoughton, 2006 [1956]), p. 143.

52 Wentworth, *The Listening Eye* (London: Hodder and Stoughton, 2006 [1957]), p. 35.

53 Ibid., pp. 4–5.

54 A. G. Crane, 'Physical Reconstruction and Vocational Education', in *The Medical Department of the United States Army in the World War*, 17 vols (Washington, DC: U.S. Government Printing Office, 1927), XIII, p. 181; Melanie W. Hudson and Mark DeRuiter (ed.), *Professional Issues in Speech-Language Pathology and Audiology*, 5th edition (San Diego: Plural, 2021), p. 59.

55 See, for instance, Ian McLaine, *Ministry of Morale: Home Front Morale and the Ministry of Information in World War II* (London: Routledge, 1979).

56 *The Next of Kin* (1942), written by Basil Bartlett, Thorold Dickinson, John Dighton and Angus MacPhail, directed by Thorold Dickinson.

57 David Karp, *One* (Richmond, VA: Valancourt Books 2017 [1953]).

58 'When meeting in public with colleagues, keep your hands, glass, or cup over your mouth to prevent lip-reading by strangers' (Richard C. S. Trahair, *Encyclopedia of Cold War Espionage, Spies, and Secret Operations* (Westport, CT: Greenwood Press, 2004), p. 430).

59 *The Listening Eye*, pp. 1–2.

60 Ibid., p. 4. In his third outing as Bond, Daniel Craig meets Ben Whishaw's Q in the National Gallery, Room 34. After reflecting on Turner's *The Fighting Temeraire*, the Quartermaster quietly informs Bond of his new mission (*Skyfall* (2012), written by Neal Purvis, Robert Wade and John Logan, directed by Sam Mendes).

61 *The Listening Eye*, p. 24. The admissibility of evidence derived from lip-reading has been much debated in the legal context: see Roderick Munday, *Evidence*, 9th edition (Oxford: Oxford University Press, 2017), p. 341.

62 *The Listening Eye*, p. 28.

63 Ibid., pp. 45, 52.

64 Ibid., p. 47.

65 Ibid., p. 46.

66 Ibid., p. 40.
67 Ibid., p. 35; Raymond Chandler, *The Long Good-bye* (London: Penguin, 2005 [1953]), p. 64.
68 Dexter, *The Silent World of Nicholas Quinn* (London: Macmillan, 1977), pp. 175, 193.
69 Ibid., p. 7.
70 Ibid., p. 18.
71 Ibid., pp. 33, 21.
72 Ibid., p. 21.
73 Ibid., p. 30.
74 Ibid., pp. 70, 33.
75 Ibid., p. 128.
76 Ibid., p. 32.
77 Ibid., p. 28.
78 Ibid., p. 36.
79 Ibid., pp. 37–8.
80 Ibid., p. 39.
81 Ibid., pp. 242–3.
82 Ibid.
83 Ibid., p. 243.
84 Agatha Christie, *The Murder of Roger Ackroyd* (London: Harper Collins, 2013 [1926]), p. 287.
85 Agatha Christie, *Murder on the Orient Express* (London: Harper Collins, 2017 [1934]), p. 263.
86 Agatha Christie, *Dumb Witness* (London: Harper Collins, 2015 [1937]), p. 314.
87 Dexter, *The Silent World of Nicholas Quinn*, pp. 243, 71.
88 Ibid., p. 249.
89 Ibid., p. 282.
90 Ibid., pp. 250, 284–5.
91 Ibid., p. 38.
92 Ibid., p. 285.
93 Ibid., p. 249.
94 Dexter, *The Jewel That Was Ours*, p. 128.
95 Colin Dexter, *Last Bus to Woodstock* (London: Macmillan, 1975), p. 242.
96 Dexter, *Death is Now My Neighbour*, p. 64.
97 Colin Dexter, *The Riddle of the Third Mile* (London: Macmillan, 1983), p. 185.
98 Ria Cheyne, *Disability, Literature, Genre: Representation and Affect in Contemporary Fiction* (Liverpool: Liverpool University Press, 2019), p. 79.
99 Ibid., pp. 3–4.
100 Wilkie Collins, *Hide and Seek*, ed. Catherine Peters (Oxford: Oxford University Press, 2009 [1854/61]); Eudora Welty, 'The Key', in *The Collected Stories of Eudora Welty* (New York: Harcourt, 1980), pp. 29–37; Barry Unsworth, *Morality Play* (New York: Anchor Books, 2017 [1995]).
101 Dexter, *The Jewel That Was Ours*, p. 236.
102 Colin Dexter, *The Remorseful Day* (London: Macmillan, 1999), p. 43.

Earpiece 2
Learning to Be Hearing

Ben Holmes

It was my first day in a new role working as a clinical psychologist in a specialist service in the NHS. All the service users and some of the professionals I would be working with were profoundly deaf and would be relying on British Sign Language (BSL) to communicate.[1] I had no prior knowledge of BSL and would be learning 'on the job'. What I had not realised until the point of walking through the door was just how Hearing I was. By this, I mean that the vast majority of people I would be working with were linguistically and culturally Deaf. I was neither.

It was several years ago now that I walked into that role. At the time, the experience left me feeling instantly deskilled and unable to communicate even the most basic things. At first, I thought that the one skill I was lacking was BSL and that this would be remedied by attending BSL lessons. Since then, I have had the privilege of learning BSL and so much more. The experience of being immersed in a world that is linguistically and culturally different to mine has helped me learn to appreciate more about language and communication in general, and about my own culture and identity – about what it means to be Hearing.

Starting to Learn

My initial memories of starting to learn BSL were of me tying my fingers in knots! I remember my lack of dexterity, trying to find ways of moving my fingers and hands into shapes they were not used to. Watching fluent signers do this effortlessly was something that amazed me. As my vocabulary increased, I found my hands continued to struggle to keep up with what I was wanting to say for quite some time.

It is commonly assumed that sign language is an analogue of spoken language, that it is a collection of signs to match each word. It is also common for people to think that all signed languages are the same around the world. Neither of these statements is true. BSL, as with other sign languages, has a grammar distinct from that of spoken English, and there is a

DOI: 10.4324/9780429291616-9

rich diversity of sign languages around the world, not to mention regional variations within each sign language (e.g. 'purple' has over 20 different signs depending on where in Britain you come from!).

Learning a new language forced me to really think about the concept I was trying to express rather than about a literal word-to-sign translation. For example, if asking 'how long?', the way I might ask would differ depending on whether I was referring to time (e.g. 'How long will it take?') or to distance (e.g. 'How long is a piece of string?!'). I also became aware of the importance of context to conversation and of how small errors can quickly lead to vastly different meanings. It is for this reason that my innocent attempts early on to describe a piece of cake to someone instead led to me swearing at them in an incredibly rude way!

Language Is So Much More Than Words (or Signs)

So many aspects of communication are non-verbal. As a psychologist, I had long had an intellectual understanding of this fact, but my experience learning BSL gave me a richer appreciation of other factors – ones that I had not really considered before – that could inhibit or enhance my communication: the visual 'noise' of my clothes, the impact of facial hair on the readability of my lip pattern, the way I position myself in relation to the person I'm communicating with, amongst others.

One of the significant differences between sign language and spoken language has to do with communicating information visually rather than orally. This may seem obvious, but it makes a fundamental difference to how ideas are communicated: the way that objects are described; the points of reference used to provide context for conversation; and what might be considered 'acceptable'. In many cultures for whom spoken language is the norm, it can be considered 'rude to point'; but if you are using a visual language, it might be essential.

Through my experiences of working with deaf people, I have come to realise the importance of these factors for all of us in all communication situations. The reality is that there is no dichotomy of deaf/hearing; there is a spectrum of experience. To varying degrees, we all use visual, aural and contextual information when communicating.

Cultural Difference

One of the first differences I became aware of that challenged me culturally was gaining someone's attention. At its simplest, this only required me to make good eye contact and maybe to wave my hand. However, for someone who is Deaf, it is also culturally normal to switch the lights on and off, to stamp on the floor, to bang a table or tap on a person's shoulder.

This gave me an insight into the 'norms' I had long taken to be the case around boundaries and personal space and actions that might be considered 'acceptable' – a link I had previously not made to language and my cultural norms.

Another personal challenge pertained to the ways I was now required to use my body and face when expressing myself in BSL. This was a linguistic necessity but one that ran contrary to many of the cultural norms I held as a Hearing person, particularly in relation to how I expressed myself at work. My views about what it meant to be professional strongly related to the words I used and to how I might minimise other forms of expression. Given this new language, how could I present myself in an emotionally neutral way? Might elements of how I was actually feeling become apparent in my sign language? Coming to understand these conflicting expectations of myself, and thinking about how to find a way of ameliorating these two positions, was a process that was both challenging and enlightening.

As my confidence in BSL grew, other cultural challenges also became apparent, all of which had a fundamental basis in language and cultural experience that I had not previously appreciated. This included differences in cultural points of reference, differences in idioms and differences in humour – although maybe this was just my inability to tell jokes!

Power

One of the assumptions commonly held by hearing people is that spoken language is a superior means of communication. This is a view that many people will hold without even realising it. However, by definition, that leads to the assumption that those who cannot speak are therefore inferior. This view is the cause of what generally leads to a power imbalance and frequently to the oppression of deaf people by hearing people.

It is not uncommon for hearing people to regard deafness as a disability, and yet so much of the difficulty that is experienced by deaf people comes from the creation of communication and language barriers that do not need to exist. Anyone can be disabled by other people using an inaccessible language – an experience I was briefly exposed to when I started my job and felt instantly deskilled by my lack of fluency in BSL.

However, it is also true that if everybody in the world used signed language rather than spoken language, this 'disability' would all but disappear. Despite many people being very interested in sign language and sympathetic to the need to make communication accessible for all, when it comes to the need for individuals to make a change to their method of communication, there is a surprising amount of resistance. This is something I have observed many times in the case of people being unwilling to book an interpreter, for example.

It is unfortunately the case that some deaf people, from early on in life, do not have access to any form of rich linguistic environment simply because others do not provide it. This can lead to language deprivation, which can have a profound impact on later language, social, cognitive and emotional functioning, and it is an experience that is entirely preventable.[2] We all have a responsibility to try and ensure this does not happen.

Shared Responsibilities

Communication between two people requires a shared responsibility to make the interaction successful. I have directly observed many first-time encounters between hearing and deaf people in which the hearing people generally appear to lose all ability to make even the most basic attempts to engage in communication: poor eye contact, mumbling to the floor, talking to anyone except the deaf person with whom they are supposed to be communicating. I have, no doubt, been one of those people in the past.

However, one of the key things that I have learned is how different language and communication can be – that there is a rich variety in the forms of expression we can use to interact and communicate with each other. Through my work, I have met some inspiring deaf people who have been willing to assume more than their share of this communicative responsibility. But through their skill at helping to put the person they are engaging with at ease, they have managed to make communication effective even when so many parts of language and culture are not shared.

Summary

I have greatly enjoyed learning BSL and passing on some of this knowledge to my family. At the start of my current role, I had no idea how much I would learn about myself, my language, my culture and my identity. These are things that I think can be hard to separate, and the links between them were something I had not appreciated before. One of the greatest gifts I have gained from this experience, however, is the rich diversity of forms of expression and understanding we can make use of. Too often, it seems that barriers exist where they do not need to; but when we make a concerted effort to engage with each other, despite any differences that may exist, good communication can still occur. To do this, a good way to start is to develop some self-awareness and to be prepared to find out what you don't know. My experience learning BSL has helped me to understand more about myself and how I interact with other people as a Hearing person.

Notes

1 People will have different preferences about the words used to refer to their hearing level, depending on many factors including identity, culture and whether their hearing level has changed over time, amongst other things. For example: deaf, Deaf, hard of hearing, hearing impaired, hearing loss. Throughout this piece I have chosen to use the word deaf unless I am specifically making a reference to people who identify as culturally Deaf (in which case I use a capital D). I recognise that this will not match the preferences of all people but do so for ease of reading rather than through intending to exclude anyone or cause offence. For further discussion, see Sally Austen and Ben Holmes, *An Introductory Guide for Professionals Working with Deaf and Hard of Hearing Clients in Clinical, Legal, Educational and Social Care Settings* ([Independently Printed], 2021).

2 See Neil S. Glickman and Wyatte C. Hall, *Language Deprivation and Deaf Mental Health* (New York: Routledge, 2019).

7 The Jabber of *Money*

Tinnitus as Metaphor and Martin Amis's Critique of Neoliberalism

Elisabeth Reichel

In a 1987 installment of his popular *Remainders* column in *The Times Literary Supplement*, Eric Korn parodies the fearsome rhetoric surrounding AIDS at the time by imagining that monetary transactions, which are transmissive of 'a deadly new ailment' and economic liberalism – with its 'financial permissiveness' – carry a lethal health risk:

> I had a dream about a deadly new ailment that was transmitted by money. The first people to go down with it were usurers and croupiers, and everyone said 'serves them right'. PAY PLAGUE: GOD'S PUNISHMENT FOR AVARICE said the posters; SERVE THE DIRTY BANKERS RIGHT said the graffiti. They changed their tune, didn't they, when they found that any exchange of goods and services, even casual labour, could permit the spread of the dreaded Surplus Labour Value, or SLV? Didn't the Government lend its backing to a campaign against excessive monetarism, pointing out that while safe earning was some help (the use of letters of credit, the so-called 'plastic johnnies'), the real protection lay in abandoning the modern heresy of financial permissiveness and the virtue of growth, and a return to the medieval values of poverty? Closed down the Stock Exchange ('a cesspit of their own making'), denounced capitalist propaganda in the media, the loonies who taught business studies to the young leafleted everybody saying the real culprit was the profit motive, didn't they?[1]

Nicky Marsh notes that this conflation of the newly fearful languages of AIDS with the sexually celebratory languages of the economic liberalism of the 1980s – clichéd as the 'sexygreedy' decade, replete with 'Big Swinging Dicks' – is representative of a counter-discourse that focuses on the risks of liberalized economic behaviour.[2] In this discourse, the viral illness serves as a metaphor that symbolically connects the excesses of unfettered free-market capitalism with death and ideas of indulgence, sexual deviance, and

DOI: 10.4324/9780429291616-10

delinquency, resulting from the minoritarian groups most severely affected by AIDS in the 1980s such as homosexual men and intravenous drug users.

No one has been more critical of such metaphorical uses of illness, and in particular AIDS, than Susan Sontag, who in *Illness as Metaphor* (1978) and *AIDS and Its Metaphors* (1989) traces linguistic patterns that tend to place blame on the patient for freely engaging in supposedly unsafe behaviours that put them at risk of contracting the illness.[3] Indeed, by imagining an AIDS-like epidemic that descends on people partaking in the new financialized economy, Korn's parody denies precisely the arbitrariness of those groups of people who come to be infected and renders their involvement in a specific form of economic organization integral to the clinical picture of the disease. Judgment is cast on 1980s neoliberalism as HIV-positive people and those living with AIDS remain symbolically linked to unsafe, excessive, and delinquent behaviors.

The present chapter sets out to analyze the use of auditory illness as metaphor in the critique of neoliberalism that Martin Amis's 1984 novel *Money: A Suicide Note* presents.[4] Yet rather than engage in a Sontagian project that interrogates the political and ethical ramifications of this rhetoric for patients outside of Amis's novel, I am more interested in the historical and discursive place of *Money*'s curious thematic coupling of auditory illness and neoliberalism critique – with Amis's protagonist, John Self, noting on the first page of the novel that '[o]wing to this fresh disease I have called tinnitus, my ears have started hearing things recently'.[5] This coupling, I propose, marks an important point in the literary history of both economic liberalism and tinnitus as the latter comes to function as a symptomatic disease (*Symptomkrankheit*) and an epochal disease (*Epochenkrankheit*) that points symptomatically and symbolically toward neoliberalism's discontents at the beginning of what current scholars of neoliberalism often describe as the Neoliberal Age.[6] Importantly, *Money*'s use of tinnitus as a symptomatic disease to critique neoliberalism enlists a humanist Enlightenment discourse that has construed 'hearing things' as indicative of a deranged mind. Rather than caution against the socioeconomic inequities that it aggravates and the social safeties that it dismantles, *Money* presents neoliberalism's new way of organizing economic relations first and foremost as a threat to the autonomy of the self and the cogito. In the process, the protagonist's disorderly hearing is rendered symptomatic of an Enlightenment subject in crisis that the novel conjures up as the product of a new economic liberalism.

Reading Martin Amis's *Money* as a refutation of Thatcherism and, more broadly, of the neoliberal policies that were implemented on both sides of the Atlantic in the early 1980s has been a commonplace in scholarly writing on the novel.[7] The particular languages that the novel adopts to conduct this critique, however, have received little sustained attention, with

Marsh's 2007 essay standing out in its careful analysis of *Money*'s use of a critical vocabulary that construes financial deregulation 'as posing a threat [. . .] to the integrity of the masculine body': in *Money*'s transatlantic world, '[t]he sexualization of money that took place in the eighties seems to have been all too easily displaced onto an anxiety about shifting gender roles and the empowerment of femininity in particular', Marsh contends.[8] Turning from gendered bodies to illness, it is also important to note that *Money* goes to great lengths – a whopping 363 pages, to be precise – to show that this new economic liberalism and financialization renders Self's body extremely unwell. Physical pain and monetary ailments mix, for instance, in an inventory of health concerns that Self takes down in a New York hotel room at the beginning of the novel: 'a definite swelling in my jaw', probably 'a fucking abscess or something', reminds him that '[t]here are taxi-meters, moneyclocks, on the ambulances in this city', until 'another pain [is] starting business in the slopes of my eyes'.[9] He is drinking tax-exempt whisky from the airport while slowly passing out to the monotonous, parallelistic tally of his debtors: 'Alec Llewellyn owes me money. Selina Street owes me money. Barry Self owes me money'. 'Refreshed by a brief blackout', he then takes stock of his physical appearance ('Jesus, I'm so fat these days.') and determines to 'fix it' through costly surgery 'the minute I hit the money': 'I need my whole body drilled down and repaired, replaced'. However, before he pays large amounts to restore his health, 'I want to go out and drink some more and get a lot tireder first', he concludes his physical examination. The excessive consumption of substances, goods, and services that leads to the complete exhaustion of Self's body is not only encouraged by the investors in his film project, his 'moneymen', who expect conspicuous consumption as a display of wealth and prosperity:

> Slick, I'm going to have to talk to you very seriously about your expenses. Shape up, John. It's an embarrassment. It looks bad to the moneymen. Take a floor at the Gustave. Hire a jet and have a weekend with Butch and Caduta in the Caribbean. Go buy a case of champagne and pour it all over your dick. Spend. Spend. You're no *use* to me when you fly coach. Fly supersonic. Fly sharp-end. God damn it, Slick, fly *right*.[10]

Even more, as Self later reveals in an argument with the character Martin Amis, he considers frugal spending habits 'immoral', a refusal to contribute one's share to society: 'You haven't got shit, have you, and how much do you earn? It's immoral. Push out some cash. Buy stuff. Consume, for Christ's sake'.[11] Given these forceful, external, and internalized pressures to perpetually '[p]ush out' money and consume 'stuff', Self can only conceive of his physical recovery in terms of more spending, a complete makeover of his body that is a 'well-earned body transplant', or, as Fielding Goodney suggests when witnessing Self's collapse during a tennis match, 'a series of long, painful and expensive operations' supplemented by regular visits to 'high-priced health clubs and fancy massage studios'.[12]

In *Money*'s portrayal of early-1980s neoliberalism, then, Self's consumerist way of life – which submits to a moral imperative to spend ever more money in an economy that follows a growth mandate – erodes his health dramatically and, indeed, appears to drive him to an early death. At one point in their game, Self imagines telling Fielding to stop '[b]ecause *I think I might* DIE if I don't'.[13] One meaning of the novel's eponymous 'suicide note' is made unequivocally clear throughout, as *Money* lays out in great detail Self's exorbitant needs: put crudely, he drinks, eats, fucks, and drugs himself to death, all the while pressing out large sums of money. 'God I can't wait to leave this moneyworld', Self blurts out when realizing that '[i]t's all money', and there is no escaping it, with even suicide coming at a price '[u]nless you're really brave'.[14] At the same time, *Money* repurposes Self's – highly expensive – death by overspending to stand in synecdochically for a global development affecting national economies on both sides of the Iron Curtain:

> Really living is what I'm doing, and it's killing me. [. . .] I subserve the detailed money action, do this, do that, run money errands. I am pussy-whipped by money, but then so is the United States. So is Russia. We are all stomped and roughed up and peed on and slammed against the wall by money. Should the earth enter turnaround tomorrow, nuke out, commit suicide, then we'll already have our suicide notes, pain notes, dolour bills – money is freedom. That's true. But freedom is money. You still need money.[15]

As Self submits to the demands of a financialized, liberalized system of economic organization, so do nation-states and political collectives, which partake in the same 'moneyworld' that lends money a totalitarian power and brutal force: 'We are all stomped and roughed up and peed on and slammed against the wall by money'. Thus, the freedom that economic liberalism promises, Self painfully realizes, is still a dependency on money, and true freedom – including freedom from money – is ultimately impossible in a neoliberal system of free markets. In this world of self-induced financial terror, Self reflects, paper currency presents the 'suicide notes, pain notes, dolour bills' that will be left '[s]hould the earth enter turnaround tomorrow, nuke out, commit suicide' and both Self's body and the entire world system collapse.

Money thus continuously interweaves, in a thematic through line, ill health and death with the new form of economic liberalism that took hold in social and cultural spheres across the Atlantic in the early 1980s. Most important to the remainder of this chapter, though, the novel suggests more specifically that this new world affects Self's auditory health in a particularly pernicious way. The 'fresh disease' that Self diagnoses on the first page as tinnitus also features prominently in the physical examination that he conducts on himself in his hotel room: 'I'm [. . .] listening to see if I'm still hearing things. The mornings are the worst. This morning was the worst yet. I heard computer fugues, Japanese jam sessions, didgeridoos'.[16] After his match with Fielding, he notes that '[t]he tinnitus was operational pretty well full time'.[17] Like the neoliberal world of money that he cannot escape, Self's tinnitus has also a transatlantic nature and follows him as he traverses different time and geographical zones: 'London is waking up. [. . .] The distant fizz or whistle or hiss in the back of my head is starting again, modulating slowly, searching for its scale'; 'the sound that lives inside my head' is 'a faint parched whistle, dreary and remote', that '[p]erhaps [. . .] was the sound the Atlantic made with all its mass and space'.[18]

While literary scholars have often centered their critical attention around the social historical contexts in which Amis's 1984 novel is embedded, the protagonist's disorderly modes of hearing have remained a largely unexamined component of its meaning-making process. The present chapter contends that it is significant that 'the most paradigmatic British novel of the fast-track greedy 1980s, and the most influential stylistically, technically, and thematically', as Elaine Showalter claims, integrates into its tangled imagery an auditory illness whose medical name was just entering popular, non-medical discourse in 1984.[19] Tinnitus is 'fresh' not only in its newness to Self's body but also as a popular descriptor.[20] In fact, Self's self-diagnosis is not without ambiguity and is often prefaced in his narration by a self-conscious reference to the naming act: 'this fresh disease I have called tinnitus'; '[t]he disease I host called tinnitus'; 'the disease I have called tinnitus'.[21] At times, the narrator-protagonist even forays into the disease's clinical picture as it emerges from new medical findings: '[e]ven when they're deaf, apparently, the ear-sufferers still have this head-noise to listen to. Bad luck. Double bad', he learns, for instance.[22] This ambiguousness of *Money*'s portrayal of tinnitus has led clinical hearing scientist David Baguley to contest Self's diagnosis and to identify 'a formless auditory hallucination' in Amis's protagonist rather than 'tinnitus as generally experienced'.[23]

In his groundbreaking cultural history, *Of Ringing Ears and Godly Voices: A Cultural History of Tinnitus* [*Ohrenrausch und Götterstimmen: Eine Kulturgeschichte des Tinnitus*] (2012), Uwe Steiner shows that, contrary to popular conception, what is today termed tinnitus dates back at least to Biblical times, with denotations and discourses around the disease having varied widely over the past centuries and millennia. Within this long cultural history, Steiner notes, tinnitus has served as a symptomatic disease (*Symptomkrankheit*) and an epochal disease (*Epochenkrankheit*), as it has been charged with additional symbolic meaning and thus rendered useful to the work of cultural criticism. Tinnitus as a symptomatic disease in this way registers cultural discontents, which become 'diskursfähig', that is, a subject of public debate.[24] By extension of these symbolizing functions, symptomatic diseases come to characterize historical periods, giving expression to larger cultural concerns and the malaises of a certain time. Studying the literary history of tinnitus, then, means tracing cultural concerns as they become manifest in the symbolic meanings that different literary texts add in their portrayal.[25]

I have proposed that Amis's *Money* avails itself of tinnitus as a symptomatic and epochal disease in its endeavor to critique a then new form of economic liberalism and financial deregulation. In a key passage of the novel, Self describes his tinnitus as 'four distinct voices', the first of which is 'the jabber of money, which might be represented as the blur on the top rung of a typewriter – £%¼@=&$! – sums, subtractions, compound terrors and greeds'.[26] By means of synaesthetic imagery, *Money* makes Self's tinnitus point to non-auditory 'things' beyond the realm of mere acoustics: 'Owing to this fresh disease I have called tinnitus, my ears have started hearing things recently, things that aren't strictly auditory'.[27] For one, he hears the calculations and emotional tolls – 'sums, subtractions, compound terrors and greeds' – of the newly liberalized and financialized economic system that he inhabits. Importantly, I further argue, the critique of neoliberalism that *Money* puts forward is grounded in an Enlightenment humanism that construes tinnitus as the index of a disordered mind and, by implication, a less advanced human subject and inferior state of civilization. *Money*'s rendering of its protagonist's hearing condition ties in with and continues a conception of 'hearing things' as 'a quintessential earmark of modern insanity'.[28]

In his encyclopedic *Hearing Things: Religion, Illusion, and the American Enlightenment* (2000), historian of religion Leigh Eric Schmidt tracks the debate over ways of listening to divine voices that erupted when auditory immediacy in Christian piety came under scrutiny by Enlightenment critics. In the process, Schmidt uncovers an important prehistory of what is today known as tinnitus: as the authority of natural philosophy widened, a shift in meaning from divine agency to perceptual fallacy and medical, institutionalized pathology took place.[29] One of the earliest comprehensive works on otology, Joseph-Guichard Du Verney's *A Treatise of the Organ of Hearing: Containing the Structure, the Uses, and the Diseases of all the Parts of the Ear* (1683) lists disorders that render 'the Ear sensible of Noises which are not in reality, or which are not external', and attributes these internal sounds to such mental states as delirium and frenzy, transitioning quickly from an account of 'the Disorders of the Ear' to 'the Disorders of the Brain'.[30] Later anatomists described causes ranging from intestinal maladies to obstructed Eustachian tubes, as well as the vague diagnosis of nervousness, which made somatic debility and mental hallucination – anatomy and psychology – entirely reciprocal. Witness Joseph Williams, for instance, in his *Treatise on the Ear; Including Its Anatomy, Physiology, and Pathology* (1840). Tinnitus, he states,

> frequently arises from an altered function of nerves, and [. . .] is much complained of by hypochondriacs, dyspeptics, and hysterical females, whose imagination sometimes converts these noises into *whispers*, proving a source of great disquiet, and amounting, in some instances, to complete hallucination.[31]

Toward the end of his treatise, Williams claims that '[l]unatics', in particular, 'are often exceedingly tormented by ringing in the ears [. . .], frequently inciting and encouraging them to self-destruction'.[32] In support of these claims, he refers back to John Haslam's *Observations on Insanity* (1798), which notes that 'maniacs often suppose they have [. . .] heard those things, which really did not exist at the time'.[33] To prevent such ideas 'grasp[ing] strongly upon the mind and seduce its credulity', Haslam recommends that the youth be taught 'to subjugate the passions, as to discipline the intellect'.[34]

The late-eighteenth- and early-nineteenth-century ascent of the asylum directly correlated with a proliferation of studies in insanity, lunacy, and mania that medicalized acoustic illusions. In the asylum's 'noiseless regimen of reason', cures for disordered hearing were to be effected through moral and intellectual habits of self-control, industry, and courtesy.[35] 'One reordering it definitely accomplished', Schmidt quips: 'hearing things became a quintessential earmark of modern insanity'.[36] While *Hearing Things* is inspired by the lasting impact of this shift in Christian piety away from spiritual practices of listening, Schmidt's findings also raise questions as to how the association of 'hearing things' with a disordered mind and deranged subjectivity – an association forged by Enlightenment anatomists and physiologists – has had a continued influence on contemporary discourses around tinnitus and the literary treatments that it receives. Martin Amis's *Money* can be read as an important testimony in this respect, as it evokes through the tinnitus-ridden perspective of its narrator-focalizer a cityscape permeated by both deregulated money and mental disorder.

Thus, apart from the voice that is 'the jabber of money', with its 'compound terrors and greeds', one of the 'subvoices' that Self identifies when taking stock of the different things he hears conjures up a disheveled figure, marked by poverty as well as by mental disarray:

> an obsessed black tramp or retard who roams the Times Square beat here in New York. Incomprehensible yet unmistakably lecherous, his gurgled monologue goes like this: Uh guh geh yuh tih ah fuh yuh uh yuh fuh ah ah yuh guh suh muh fuh cuh. I do a lot of that kind of talking in my head too.[37]

Coalescing monetary and mental precariousness, and connecting Self with both, the figure forebodes the disastrous consequences that Self's liberalized way of life risks – that is, financial ruin and loss of sanity – with the latter being prefigured by the '[i]ncomprehensible', 'lecherous', 'gurgled monologue' that Self utters 'in [his] head too'. A similarly ill-boding figure, who 'loosely babbled of fraud and betrayal, redundancy, eviction', appears in one of the novel's first paragraphs:

> So I walked south down bending Broadway. What's all this *mmhm* shit? [. . .] I heard the ragged hoot of sirens, the whistles of two-wheelers and skateboarders, pogoists, gocarters, windsurfers. I saw the barrelling cars and cabs, shoved on by the power of their horns. [. . .] Urged out from the line of shufflers and idlers, watchers, pavement men, a big blond screamer flailed at the kerb, denouncing all traffic. His hair was that special mad yellow, like an omelette, a rug omelette. As he shadow-boxed he loosely babbled of fraud and betrayal, redundancy, eviction. 'It's my money and I want it!' he said. 'I want my money and I want it now!' The city is full of these guys, these guys and dolls who bawl and holler and weep about bad luck all the hours there are. I read in a magazine somewhere that they're chronics from the municipal madhouses. They got let out when money went wrong ten years ago . . . Now there's a good joke, a global one, cracked by money. An Arab hikes his zipper in the sheep-pen, gazes contentedly across the stall and says, 'Hey, Basim. Let's hike oil.' Ten years later a big whiteman windmills his arms on Broadway, for all to see.[38]

The 'big blond screamer' evokes both mental and financial ruin, and barely sticks out in a New York soundscape full of sirens, whistles, and horns as well as 'guys and dolls who bawl and holler and weep about bad luck all the hours there are'. Given what readers have already learned about Self's new tendency to hear things that are not there and his unreliability as a narrator-focalizer, 'this *mmhm* shit' may well belong to the internal soundscape that his tinnitus creates, since the boundary to any external referent of Self's acoustic perception has been destabilized. Indeed, it is also unclear where 'these guys and dolls who bawl and holler and weep about bad luck' precisely come from and pertain to. Self claims to have 'read in a magazine somewhere' that they are 'chronics from the municipal madhouses' who were let out 'when money went wrong ten years ago'. The instability of the referent of Self's perception – what is wrong with him in 1981 – is related, the novel implies, to what 'went wrong' with money a decade earlier, leading to mad people walking the streets of Midtown Manhattan, including the protagonist himself. In 1971, the United States effectively ended the Bretton Woods agreement of 1944 by no longer guaranteeing the convertibility of the US dollar into gold (at 35 USD per ounce). The value of the dollar would be determined, then, in subsequent years on global financial markets, by perceptions of its worth in relation to other currencies. With characteristic historical acuity, Marsh thus rightly observes that Amis conflates two different historical events when referring at the end of this passage to the 1973 'hike' in oil prices following the OPEC embargo.[39] However, even though the latter did not take place ten years prior to the action of *Money*, by blending these events, the novel submits that it was precipitated by the termination of the Bretton Woods system, with global prizes turning into a 'joke, [. . .] cracked by money', as they lost their reference to the real value of a material object. In the new, post-Bretton Woods financial system, monetary values can fluctuate abruptly and arbitrarily, at the random pleasure of a faraway trader: note the sexual innuendo when, 'contentedly', the 'Arab' that Self imagines 'hikes his zipper' before suggesting 'Let's hike oil' to his companion in the stall.

After the end of Bretton Woods, Marsh notes in her most recent book, 'paper appeared to supersede gold, abstraction to supersede materiality, perception to supersede the real and the volatile and short-term desires of markets to supersede the long-term planning of sovereign states'.[40] This is the 'moneyworld' from which Self is unable to escape, as money only signifies more money without ever referring to a material reality, and the signified is continually deferred.[41] Taking the end of the Bretton Woods agreement as a 'lapsarian rupture' in which the referent of international currencies was lost, *Money* unfolds a narrative that, as described by Marsh, 'was to accord especially well with literary and cultural critics of the 1980s and 1990s for whom it was analogous to culture's own accelerated processes of abstraction'.[42] Jean Baudrillard's conception of 'power' that 'floats like money, like language, like theory' informed a generation of critics for whom money in the post-Bretton Woods era, 'which promises nothing but an identical copy of itself', meant further confirmation of the loss of access to the referent in postmodernity.[43] Much of the New Economic Criticism, including works by Jean-Joseph Goux, Brian Rotman, and Marc Shell, hinges on claims of a structural homology between money and linguistic signs and on the theoretical and cultural implications of this homology.[44] Marsh also identifies 'causal parallels between the severing of paper from gold, fiction from history and cause from effect' in the writings of Fredric Jameson and David Harvey on postmodernism and postmodernity.[45] Harvey, for instance, contends that '[t]he breakdown of money as a secure means of representing value has itself created a crisis of representation in advanced capitalism', as 'the de-linking of the financial system from active production and from any material monetary base calls into question the reliability of the basic mechanism whereby value is supposed to be represented'.[46]

Clearly, *Money*, which is set in the same year that Baudrillard's *Simulacra and Simulation* appeared, is well aware of such lines of thought. Marsh reads the novel as 'a precise intervention within this debate' and more specifically as 'a self-conscious conversation with the reader regarding the forms of storytelling that can most appropriately engage with the telling parallels between finance capital's disregard for the real and literary language's disregard for referentiality'.[47] The novel's treatment of tinnitus, I would add, is integral to its engagement with the question of the relevance of these parallels as well. Tinnitus as portrayed by *Money* generates an overwhelming mass of acoustic signifiers that lack a referent in material reality. Like the dollar signs in which Self communicates as a liberalized economic subject, the signs that his disordered sense of hearing produces are not grounded in an object other than themselves and the unsteady perceptions of people.

While the novel goes to great lengths, then, to show the 'earache and riot' on its protagonist's mind and 'the money chaos' with which the city 'crackle[s]', the 'new' problem that is the root cause of this anguish, *Money* claims, lies in an absence that Self cannot quite put a finger on:

> Something is missing from the present too. Wouldn't you say? Mobile, spangled and glamorous, my life looks good – on paper, anyhow – but I think we're all agreed that I have a problem. Not so? Then what is it? Brother, sister, do the right thing here and let me in on it. Help me out. You'll tell me it's the booze . . . the booze isn't brill, I warrant, but the booze is nothing new. Something else is new. I feel invaded, duped, fucked around. I hear strange voices and speak in strange tongues. I get thoughts that are way over my head. I feel violated . . . [. . .] I sit here defencelessly, my mind full of earache and riot. Why? Tell. Inner-cities crackle with the money chaos – but I've *got* money, plenty of it, I'm due to make lots more. What's missing? What the hell else is there?[48]

When considered 'on paper', Self's life 'looks good': '[m]obile, spangled and glamorous', it lavishly displays signifiers of wealth and prosperity, with only his alcoholism – which 'isn't brill[iant]' – disturbing this picture of plenitude and success. Nevertheless, he cannot help but feel a distinct sense of loss. This is a new ailment that cannot be connected to his much older drinking habits; nor will more money solve this issue, since '[he's] *got* money, plenty of it,' and '[he's] due to make lots more'. Searching for something to fill this void, Self is '[r]andomly prompted', he claims, to peruse his 'book collection' – which is essentially a pileup of titles that his former girlfriends did not bother to pick up after they left: '*Home Tax Guide*, *Treasure Island*, *The Usurers*, *Timon of Athens*, *Consortium*, *Our Mutual Friend*, *Buy Buy Buy*, *Silas Marner*, *Success!*, *The Pardoner's Tale*, *Confessions of a Bailiff*, *The Diamond as Big as the Ritz*, *The Amethyst Inheritance* – and that's about it'.[49] He then moves on to his 'space-age sound system', which reminds him that he lost interest in music a long time ago, leaving him ultimately only with '[w]atching television' as 'one of my main interests, one of my chief skills', and video films as 'another accomplishment'. That is, rather than '[r]andomly prompted', Self's search for something that alleviates his sense of loss is informed by Enlightenment ideas about the meaning of human life, chief among which is the notion of a linear progression toward a supremely refined, civilized state of development through intellectual achievements and an education in the arts. Being interested in neither books nor music, though, Self's 'accomplishment[s]' in this sense are limited to watching TV and film, specifically 'diabolism, carnage, soft core'.

In fact, 'when [he] can bear to think about it', none of his interests adds meaning to his life in such a way as to elevate humankind; instead, they are all 'pornographic in tendency': 'The element of lone gratification is bluntly stressed'. This realization prompts another attempt on Self's part to determine a stable referent that alleviates the lack that he senses: his sudden 'need' for the 'human touch' of another person, which is 'free, complimentary, with no cash attaching'.[50] Tired of the arbitrary language of dollar notes in which he conducts his financial transactions, '[j]ust once, after all these years' he wishes to communicate through love notes, which use meaningful signs to refer to the beloved:

> I want to say, Look, can't we change the subject? [. . .] Isn't there anything else you can talk about? . . . When did I last get a love letter, for Christ's sake? When did I last write one?'

It is in this desperation for true affection, 'looking for a friendly, a helping hand', that he goes through his mail and discovers a 'human touch' on a hand-addressed letter. Tellingly, he first suspects this gesture of being part of a marketing ploy that imitates a language of love to make him spend more money, as for instance at a hairdresser's or a massage parlour: 'They get foreign chicks to hand-address the mail: it lends a personal touch'. The letter is from his lover, Selina Street, however, who also uses signs of affection to entice Self to provide her with financial means: 'PS. I'm pennyless', the note ends. *Money* goes on to attend to Selina for most of its narration to illustrate the extent of Self's 'pornographic' tendencies, his penchant for pleasure-seeking involving monetary transactions.[51] Although Selina manages to renegotiate the terms of their relationship, thereby increasing her standing from being the recipient of physical violence to holding a joint bank account and warding off some of Self's advances, her role remains pornographic as she sexually gratifies Self at his will and under the pressure of his financial hold over her:

> Selina Street has no money, no money at all. [. . .] She has stolen. She has pawned clothes. She has fucked for money. [. . .] She has always said that men use money to dominate women. I have always agreed. That's why I've never wanted to give her any.[52]

The novel's extended portrayal of these power dynamics as played out in the sexual relations of Self and Selina is often gratuitous in its detail. Clearly, their relationship is far removed from the romantic ideal of unconditional love and 'human touch' that prompted Self's search – 'free, complimentary, with no cash attaching' – and fails to provide the non-monetary, stable referent that is missing from his life.[53] *Money* contrasts this transactional, pornographic relationship sharply with Self's short-lived romance with his college crush, Martina Twain, who, as opposed to Selina, 'has always had money – she has never not had money', and is both 'knowing' and 'innocent, because money makes you innocent when it's been there all along'.[54] For Self, Martina is the epitome of 'class, with a terrific education on her' and a cosmopolitan way of life: 'The long legs have travelled, and not just through time. The clean tongue speaks French, Italian, German. The expectant eyes have seen things, and expect to see more'.[55] Her freedom from the monetary, 'world[ly]' pressures that torment most of Self's contemporaries has the effect that, as Self puts it in an address to the reader, 'she sounds sane, doesn't she, among all these other people I'm working around?'. Sanity, in the new, neoliberal world that Amis's protagonist and 'all these other people [he's] working around' inhabit, is a luxury that only the comfortably rich can maintain. Given that *Money* takes Self's tinnitus to be symptomatic of the mental disorder resulting from this neoliberal form of economic organization, it is no surprise that Self feels his 'head-static quieten' when he spends time with Martina or thinks of her: 'the thing about Martina is that I can't find a voice to summon her with. [. . .] I think of her and there is speechless upheaval in me'.[56]

When Self and Martina engage in a romantic relationship toward the end of the novel, then, *Money* interjects in its tragic narrative a brief, retarding moment of hope for an alternative conclusion other than catastrophe. This moment again suggests the return to an older, Enlightenment humanist form of subjectivity as the cure for the issues that neoliberalism and financial deregulation cause in both Self's auditory and mental health. From her privileged position where money has always been 'carelessly present', affording her education and leisure, Martina takes it upon herself to enlighten Self and elevate him, in a civilizing mission, to a literate, cultured member of human society.[57] Thus, Self turns from someone who 'do[es]n't know what it's like to write a poem' and 'do[es]n't know what it's like to read one either' and who thinks of reading as a source of pain[58] into a dedicated reader of books that Martina provides, covering key figures of a Western canon: '*Freud, Marx, Darwin, Einstein* and *Hitler*'.[59] The benefits of '[t]he bookish, the contemplative life' for his auditory and mental health are immense, Self notes:

> Martina, she's even cured my tinnitus. Not a squeak for over three hours. The big thing about reading and all that is – you have to be in a fit state for it. Calm. Not picked on. You have to be able to hear your own thoughts, without interference.[60]

This recovery in mental fitness is accompanied by an enhanced ability to make meaning of his surroundings ('I could make a little more sense of the watchers and the watched.') as monetary values cease to function as Self's primary system of communication and are replaced by non-monetary sign systems. The first book that Self receives from Martina, George Orwell's *Animal Farm*, which later turns out to be a library book, is the first 'proper present' that he has received in a very long time.[61] Self's life also appears to take on new meanings now that literature has become his primary mode of sense-making. After reading Orwell's *1984*, he reports,

> I saw with a jolt that the room I had hired was Room 101. Perhaps there are other bits of my life that would take on content, take on shadow, if only I read more and thought less about money.[62]

However, it is at this stage of mere reference that Self's hermeneutical skills end. His inability to interpret texts beyond their first, literal reading is one of the novel's great sources of humour. Missing its allegorical meanings entirely, Self relates to *Animal Farm* in a profound manner as a story about different animals. He ponders,

> Where would *I* be in *Animal Farm*? [. . .] Now, after mature consideration, I think I might have what it takes to be a dog. I *am* a dog. [. . .] I am bouncing, twisting, weeping, consuming myself. [. . .] Look at the

> dogs in the street, how everything implicates them, how everything is their concern, how they race towards great discoveries. And imagine the grief, tethered to a fence when there is activity – and play, and thought and fascination – just beyond the holding rope.[63]

Money takes this animal imagery further in its parallel portrayal of Martina's rescue of a dog and Self's momentary redemption under her care. Martina's nurturing of her ailing lover is echoed in the domestication of her stray dog Shadow, whom she adopts around the same time that she becomes involved with Self romantically and who takes to the latter 'at once, of course, as dogs do'.[64] The failure of this civilizing endeavour is thus also foreshadowed in Shadow's animal behaviour and the daily struggle to keep him from scampering back to his old hunting ground, 'the sin and death of Twenty-Third Street, Chelsea, the world's end, where everything was unleashed, unmuzzled. There were no collars, no leads, no names on Twenty-Third Street'.[65] Self tries to reassure Martina that Shadow 'knows what the good life is', and, by extension, so does he.[66] Yet the novel ultimately validates her concerns that 'it's his *nature*' to live unleashed, unmuzzled, with both Self and Shadow 'answering to a sharper nature' when they finally free themselves of Martina's – literal and figurative – leash.[67] Gone is 'the grief', as Self anthropomorphizes the dog's feelings in his reading of *Animal Farm*, 'when there is activity – and play, and thought and fascination – just beyond the holding rope'.[68]

After being elevated by Martina to a higher, civilized stage of human development, Self's betrayal of her not only coincides with Shadow's final escape.[69] As James Diedrick notes, it is also 'accompanied by a symbolic devolution' in which Self takes on an insectile form: Selina calls him as he is holding an aerobic position 'wearing tacklebag only and wiggling my legs in the air like an upended beetle', 'swearing and mumbling and spitting the fluff from my mouth'.[70] When Martina later bursts in on them mid-intercourse, he finds himself in a similar insect-like position, 'flummoxed, scuppered, and waving his arms'.[71] Alluding to Gregor Samsa's transformation in Kafka's *Metamorphosis*, Self devolves from his 'new-deal me' to the subhuman state of a beetle – or to that of an '*inhuman dog*', which is the phrase from *Othello* that Fielding uses to put Self down.[72]

In taking, then, the failure of Martina's civilizing mission and its promise of a cure for Self's disordered mind and hearing to be the necessary result of a primitive, animalistic nature that he possesses, *Money* renders its catastrophic end inevitable. In the naturalistic logic of the novel, Self cannot help but fall back into his unhealthy ways because of an innate desire to live 'unleashed, unmuzzled'.[73] This is critical to the novel's conception of economic liberalism and liberalist forms of economic organization, which thus emerge as a more natural habitat for Self, answering to a core desire for unregulated freedom. *Money*'s critique of neoliberalism in this way prophesies nothing short of an inevitable downfall of human society, its 'suicide' to be specific. For at the same time, *Money* insists persistently, the new environment of financial permissiveness, with its impenetrable language of monetary signs that have no fixed referent, also gives rise to disease and mental decay. I have argued that tinnitus, the 'fresh disease', is thus rendered symptomatic of a new, neoliberal age.[74] In the process, *Money* enlists an Enlightenment humanist discourse that takes hearing things to be a marker of modern insanity, forging a strong symbolic link between neoliberal subjectivity and a cogito that is dangerously out of order.

Amis's 1984 novel antedates more recent critiques of neoliberalism that also tap into humanist rhetoric when considering the forms of subjectivity and the particular 'rationality' that neoliberalism has generated over the past decades.[75] Wendy Brown's *Undoing the Demos: Neoliberalism's Stealth Revolution* (2015), arguably the most influential exegesis of Michel Foucault's lectures on neoliberalism, resonates in interesting ways with the case that *Money* makes; it does so by arguing that neoliberalism renders ubiquitous the stunted form of human existence which Aristotle and Hannah Arendt described as 'mere life' and Karl Marx as life 'confined by necessity' – that is, life limited to survival and wealth acquisition.[76] Neoliberalism, according to Brown, eliminates 'the cultivation and expression of distinctly human capacities for ethical and political freedom, creativity, unbounded reflection, or invention' – that is, 'the good life' (Aristotle) and 'the true realm of freedom' (Marx).[77] In other words, '[n]eoliberalism is the rationality through which capitalism finally swallows humanity', Brown surmises.[78] *Money*'s use of tinnitus as a symptomatic disease that points to and stigmatizes a similar dehumanizing development under neoliberalism is significant, I venture. It speaks to the importance of auditory illness as metaphor in literary meaning-making processes and their critical functions in relation to socioeconomic shifts. A comprehensive understanding of these processes requires from the medical humanities an analytical scrutiny that is informed by literary and cultural histories of sound and modes of acoustic perception. It is to the larger goal of expanding these histories that the present chapter has sought to make a contribution.

Notes

1 Eric Korn, 'Remainders', *The Times Literary Supplement*, 4381 (20 March 1987), 296.

2 Caryl Churchill, *Serious Money* (London: Methuen, 1987); Michael Lewis, *Liar's Poker* (London: Coronet, 1989); Nicky Marsh, 'Taking the Maggie:

Money, Sovereignty, and Masculinity in British Fiction of the Eighties', *Modern Fiction Studies*, 53.4 (2007), 845–66 (846).

3 Susan Sontag, *Illness as Metaphor* (New York: Farrar, Straus and Giroux, 1978); Susan Sontag, *AIDS and Its Metaphors* (New York: Farrar, Straus and Giroux, 1989).

4 Martin Amis, *Money: A Suicide Note* (New York: Viking, 1984).

5 Ibid., p. 7.

6 Uwe C. Steiner, *Ohrenrausch und Götterstimmen: Eine Kulturgeschichte des Tinnitus* (München: Fink, 2012). The term 'Neoliberal Age' is used by current scholars of neoliberalism that follow David Harvey in taking the late 1970s as a revolutionary turning point in world history, with Margaret Thatcher in Britain, Ronald Reagan in the United States, and Deng Xiaoping in China all beginning to implement major policy changes in an effort to liberalize economic markets. See, for instance: Rachel Greenwald Smith, *Affect and American Literature in the Age of Neoliberalism* (Cambridge: Cambridge University Press, 2015); Mitchum Huehls, *After Critique: Twenty-First-Century Fiction in a Neoliberal Age* (Oxford: Oxford University Press, 2016); David Harvey, *A Brief History of Neoliberalism* (Oxford: Oxford University Press, 2005).

7 See, for example: James J. Miracky, 'Hope Lost or Hyped Lust? Gendered Representations of 1980s Britain in Margaret Drabble's *The Radiant Way* and Martin Amis's *Money*', *Critique*, 44.2 (2003), 136–43; Laura L. Doan, '"Sexy Greedy *Is* the Late Eighties": Power Systems in Amis's *Money* and Churchill's *Serious Money*', *Minnesota Review*, 34–35 (1990), 69–80; and Dominic Head, *The Cambridge Introduction to Modern British Fiction* (Cambridge: Cambridge University Press, 2002), especially chapter 1.

8 Marsh, 'Taking the Maggie', p. 863.

9 Amis, *Money*, pp. 10–11.

10 Ibid., p. 169.

11 Ibid., p. 243.

12 Ibid., pp. 22, 39.

13 Ibid., p. 37.

14 Ibid., pp. 335, 334.

15 Ibid., p. 250.

16 Ibid., pp. 7, 10–11.

17 Ibid., p. 39.

18 Ibid., pp. 16, 31.

19 Elaine Showalter, 'Ladlit', in *On Modern British Fiction*, ed. Zachary Leader (Oxford: Oxford University Press, 2002), pp. 60–76 (66); for details of tinnitus's discursive history, see Steiner, *Ohrenrausch und Götterstimmen*, pp. 182–4.

20 Amis, *Money*, p. 7.

21 Ibid., pp. 7, 16, 96.

22 Ibid., p. 217.

23 David M. Baguley, 'Tinnitus and Hyperacusis in Literature, Film, and Music', in *Tinnitus: Clinical and Research Perspectives*, ed. David M. Baguley and Marc Fagelson (San Diego: Plural, 2016), pp. 1–11 (2).

24 Steiner, *Ohrenrausch und Götterstimmen*, p. 13.

25 Ibid., pp. 17–18.

26 Amis, *Money*, p. 104.
27 Ibid., p. 7.
28 Leigh Eric Schmidt, *Hearing Things: Religion, Illusion, and the American Enlightenment* (Cambridge, MA: Harvard University Press, 2000), p. 198.
29 Ibid., pp. 188–98.
30 Joseph-Guichard Du Verney, *A Treatise of the Organ of Hearing: Containing the Structure, the Uses, and the Diseases of All the Parts of the Ear* (London: Printed for Samuel Baker, 1737 [1683]), pp. 135–45.
31 Joseph Williams, *Treatise on the Ear, Including Its Anatomy, Physiology, and Pathology* (London: John Churchill, 1840), p. 206.
32 Ibid., pp. 250–1.
33 John Haslam, *Observations on Insanity: With Practical Remarks on the Disease, and an Account of the Morbid Appearances on Dissection* (London: Printed for F. and C. Rivington, 1798), p. 5.
34 Ibid., pp. 11, 101.
35 Schmidt, *Hearing Things*, p. 192.
36 Ibid., p. 198.
37 Amis, *Money*, p. 104.
38 Ibid., p. 12.
39 Nicky Marsh, '*Money*'s Doubles: Reading, Fiction and Finance Capital', *Textual Practice*, 26.1 (2012), 115–33 (116).
40 Nicky Marsh, *Credit Culture: The Politics of Money in the American Novel of the 1970s* (Cambridge: Cambridge University Press, 2020), p. 1.
41 Amis, *Money*, p. 335.
42 Marsh, *Credit Culture*, p. 2.
43 Jean Baudrillard, *Simulacra and Simulation*, trans. Sheila Faria Glaser (Ann Arbor: University of Michigan Press, 1994 [1981]), p. 24; Brian Rotman, *Signifying Nothing: The Semiotics of Zero* (London: Macmillan, 1987), p. 5.
44 For example, Jean-Joseph Goux, *The Coiners of Language* (Norman: University of Oklahoma Press, 1994); Brian Rotman, *Signifying Nothing*; Marc Shell, *The Economy of Literature* (Baltimore: Johns Hopkins University Press, 1978); Marc Shell, *Money, Language, and Thought: Literary and Philosophic Economies from the Medieval to the Modern Era* (Berkeley: University of California Press, 1982). The movement culminated in *The New Economic Criticism: Studies at the Intersection of Literature and Economics*, ed. Martha Woodmansee and Mark Osteen (London: Routledge, 1999).
45 Marsh, *Credit Culture*, pp. 1–2; Marsh, '*Money*'s Doubles', p. 118. Fredric Jameson, *Postmodernism, or, the Cultural Logic of Late Capitalism* (London: Verso, 1991); David Harvey, *The Condition of Postmodernity: An Enquiry into the Origins of Cultural Change* (Cambridge, MA: Blackwell, 1989).
46 Harvey, *Condition of Postmodernity*, pp. 297–8.
47 Marsh, '*Money*'s Doubles', p. 119.
48 Amis, *Money*, p. 67.
49 Ibid.
50 Ibid., p. 68.
51 Apart from '[t]he element of lone gratification' (ibid.), pornography, for Self, is defined by a transactional element: 'I don't know how to define

pornography – but money is in the picture somewhere. There has to be money involved, at one end or the other. Money is always involved' (ibid., p. 291).

52 Ibid., p. 88.

53 Ibid., p. 68.

54 Ibid., p. 128.

55 Ibid., pp. 41, 128.

56 Ibid., pp. 41, 114.

57 Ibid., p. 128.

58 'I can't read because it hurts my eyes. I can't wear glasses because it hurts my nose. I can't wear contacts because it hurts my nerves. So you see, it all came down to a choice between pain and not reading. I chose not reading' (ibid., p. 44).

59 Ibid., p. 263.

60 Ibid., p. 130.

61 Ibid., pp. 152, 130.

62 Ibid., p. 207.

63 Ibid., p. 193.

64 Ibid., p. 264.

65 Ibid., p. 267.

66 Ibid., p. 268.

67 Ibid., pp. 267–8.

68 Ibid., p. 193.

69 Ibid., p. 320.

70 James Diedrick, *Understanding Martin Amis* (Columbia: University of South Carolina Press, 2004), p. 94; Amis, *Money*, pp. 312, 313.

71 Amis, *Money*, p. 319.

72 Franz Kafka, *Metamorphosis*, in *Metamorphosis and Other Stories*, trans. Michael Hofmann (London: Penguin, 2007), pp. 73–126; Amis, *Money*, p. 312; ibid., p. 347. What Self mishears at the time as 'Oh damn dear go [. . .] Oh and you man dog' (ibid., p. 322) is later revealed to be Roderigo's line in *Othello*: 'Oh damn'd Iago! Oh inhuman dog!' (V.i.69).

73 Amis, *Money*, p. 267.

74 Ibid., p. 7.

75 Michel Foucault, *The Birth of Biopolitics: Lectures at the College de France, 1978–79*, ed. Michel Senellart, trans. Graham Murchell (Basingstoke: Palgrave Macmillan, 2008), p. 3 et passim.

76 Wendy Brown, *Undoing the Demos: Neoliberalism's Stealth Revolution* (New York: Zone, 2015); Hannah Arendt, *The Human Condition* (Chicago: University of Chicago Press, 1958); Aristotle, *Politics*, trans. C. D. C. Reeve (Indianapolis: Hackett, 1998); Karl Marx, 'On the Realm of Necessity and the Realm of Freedom', from *Capital* (III), in *The Marx-Engels Reader*, ed. Robert C. Tucker (New York: Norton, 1978), pp. 439–42.

77 Brown, *Undoing the Demos*, p. 43.

78 Ibid., p. 44.

8 Sound Minds

Schizophonia and Schizophrenia in David Foster Wallace's *Infinite Jest*

William Allen

Sound carries strangely in David Foster Wallace's 1996 novel, *Infinite Jest*. The plot pivots on 'Infinite Jest', an entertainment cartridge so compulsively entertaining that anyone who starts watching it dies of dehydration in front of the television. The film seems like a profoundly aural project. A 'wraith' incarnation of its creator, James Incandenza, says his intention was to 'contrive a medium via which he' and his son could 'simply converse [. . .] A way to say I AM SO VERY, VERY SORRY and have it *heard*.'[1] But he's in the wrong game. James is a 'cutting-edge optics man', a 'brilliant optician and technician,' but 'an amateur at any kind of real communication'.[2] And the film, according to its solo star, turns out as 'a failed piece of art'.[3] This chapter will argue that treating *Infinite Jest* like 'Infinite Jest', as something to be *heard*, clarifies its preoccupations with psychiatrist R. D. Laing's *The Divided Self* (1960), a text which attempts 'an existential-phenomenological account of some schizoid and schizophrenic persons'.[4] Wallace owned and annotated a copy, and I will argue that *Infinite Jest* recreates Laing's phenomenology of the 'schizoid' divided self within the syllabic soundscapes of sentences, using repetition, rhyme, chiasmus, and alliteration to embed echoing patterns of entombing 'in-twos' into paragraphs. After establishing Garrett Stewart's model of phonemic reading as a critical earpiece, this chapter will move back and forth between misappropriations of schizophrenia in twentieth-century critical theory, and moments of aural entombment in *Infinite Jest*. I will argue that the soundscapes of the selected sections transcend stylistic effect. Rather, the aural animation of Laing's writing – in claustrophobic, echolalic set-pieces – innovates upon some of postmodern theory's most influential but inadequate representations of schizophrenia. These, as we shall see, elide the lived experience of the illness. Stephen J. Burn writes that *Infinite Jest* has 'a specific preoccupation with forcing the reader "to learn how schizophrenia manifested itself in the human body's brain"'.[5] I will suggest an auditory counterpoint: that the text makes the reader *hear* how schizophrenia manifests itself.

DOI: 10.4324/9780429291616-11

Wallace records his own receptiveness to the aural potential of literary texts in his correspondence with Don DeLillo. Having acknowledged in one 1996 letter that DeLillo's comments on 'the physical architecture of clauses and words and letters' were 'real interesting' – by 'physical architecture' he means 'the actual appearance of words on a page', right down to 'letter-shapes and letter-combinations' – Wallace admits that he 'identified with them not one whit', since he considered himself 'maybe 100% aural'.[6] Another letter praises how one must 'read DeLillo-prose aurally, musically – less scanning than trying to feel the thump and stress', celebrating the 'rhythmic surge' of DeLillo's 'alliteration and assonance'.[7] Perhaps praise wavers into projection, and some critics have identified a similar density of sonic patterning in Wallace's work. Tom LeClair hears *something* in *Infinite Jest* and names the 'inherent disorder of dynamical systems and the neurological noise of mental illness' as two of the novel's 'conceptual bases'.[8] Burn has stressed that much of Wallace's fiction was 'written with its aural dimension prioritised' and analyses the 'aural effects' and 'alliterative structures' in Wallace's 1999 'A Radically Condensed History of Postindustrial Life'.[9] Mary K. Holland zooms in on the 'perfect aural tactility' of another phrase in the same collection, noting the 'bland a's and ah's amid brutal consonant clusters' in '"blank slack gagged masks' mindless stare"'. She concludes that Wallace was 'in moments too seldom discussed a master of poetic language'.[10] More recently, Mary Shapiro has taken this seldomness seriously, hearing these critical instances of aural alertness as exceptions. She argues that critics have emphasised Wallace's 'interest in the *philosophy* of language' over the 'phonetics and phonology' of specific sentences, and writes that there has been 'relatively little criticism of his work informed by linguistic analysis'.[11]

Taking Holland 'up on her challenge', Shapiro plumbs 'the depths' of '"blank slack gagged masks' mindless stare"', charting the space between the words on the page and the sounds articulated in the reader's brain and body. Reading the sentence, the reader is 'forced not just to *imagine* gagging, but literally to re-enact it in the repeated velar stop consonants – and in response, we hiss'.[12] Shapiro also issues a challenge of her own when she writes that 'with the rise of cognitive poetics adding science and substance to reader response theories, questions of intentionality seem much less important now than questions of mental states activated and altered within readers'.[13] I will argue that there's an audible overlap between how *Infinite Jest* represents and activates particular mental states. The virtual soundscapes of the storyworld sometimes syncopate with the verbal soundscapes of sentences, and the text's aural dimensions become charged with representing *and eliciting* altered modes of affect and cognition. Steven Connor notes 'the privileged role' hearing plays in 'the expression and experience of feeling', which 'has come to be associated overwhelmingly with auditory and acoustic experience'. Cognitive operations, meanwhile, like 'thinking, knowing, reflection and judgement [. . .] are experienced in terms of the eye'.[14] This dichotomy plays through *Infinite Jest*. Sound solidifies and sharpens in sensory, affective ways, figuring frequently in varied feelings of intensity, good or bad. Kate Gompert describes her experience of depression in auditory terms: 'everything sounds harsh, spiny [. . .] like every sound you hear [. . .] has teeth'.[15] As with James Incandenza's 'Infinite Jest', lenses muddle the potential for empathetic connection: Kate's doctor's 'eyes appeared keenly interested,' but 'in an abstract way [. . .] behind his attractive but thick glasses'; while his other patients complain of feeling 'like something in a jar he was studying intently through all that thick glass'.[16]

Sound might get through to people more than sight. As Ruth Salvaggio observes, sight always 'entails some distancing between observer and observed', while aural environments involve 'immersion'.[17] This immersion is not necessarily nice, and *Infinite Jest*'s soundscapes repeatedly play up. So shot through are they with disembodied and hallucinated noises that the text takes on the aural topography of a haunted house; it makes its reader and its characters hear things. Testing the mic for her radio show, Madame Psychosis 'makes some little sounds, little plosives to check for roaring sounds'. Little plosives follow literally, as the text tells us this is a 'perennial problem in low-budget FM'. She then makes a sibilantly 'long-*s* sound', and 'the student engineer takes a hit from his portable inhaler'.[18] Elsewhere, the same nerviness toward aural phenomena that afflicts Avril Incandenza, who 'has some auditory thing about broadcast sound and gets the howling fantods from any voice that does not exit a living corporeal head', infiltrates the text.[19] What gives Avril the howling fantods, specifically, is schizophonic sound. Composer and music theorist R. Murray Schafer coined 'schizophonia' in 1969 to refer 'to the split between an original sound and its electroacoustical transmission or reproduction.'[20] It can cover any sound that is recorded and replayed, as well as acousmatic sounds without identifiable sources, whether piped Muzak in shopping centres or Basil Fawlty sat navs.

It's fitting that a composer's coinage should depend so much on its sound. The word's strict meaning comes from Schafer's splicing-together of two ancient Greek roots: *σχίζειν* [schizein], meaning 'to split'; and *φωνή* [phoné], which means 'voice'. However, its homophonic echo of schizophrenia makes us hear double. Indeed, Schafer meant schizophonia to be 'a nervous word. Related to schizophrenia', with 'the same sense of aberration and drama'.[21] The coupling is not comfortable. The pun piggybacks on popular misconceptions of the illness in order to twist artificial implications of fear and dread into itself, and the diction of 'aberration and drama' is flippant and misleading. As psychiatrist Paul Crichton writes,

> the rise of the term 'schizophrenia' was probably due [. . .] to the simultaneously exotic and familiar sound of the word itself, with its echoes of 'schism' and 'frenzy' and 'frenetic'. [This] bears virtually no resemblance to schizophrenia as currently diagnosed.[22]

However, schizophonia in *Infinite Jest* is frequently fearful. Over 1,079 pages, there are exactly one hundred separate 'squeaks', in a narrative in which 'to hear the squeak' is a 'euphemistic locution [. . .] for instant, terrifying, and violent death'.[23] Some squeaks are real, like the 'squeaks, slow and soft but not stealthy [. . .] of weighted wheelchairs' Lucien Antitoi hears without seeing, before being killed by an elite cell of wheelchair-using assassins.[24] Others are hallucinated, like the bits of bed an unnamed neurotic father hears 'gibber and squeak' one page later, like 'hordes of gibbering squeaking rapacious rats'.[25] These squeaky moments are doubly schizophonic: their sources are suitably absent or invisible, but they also represent sounds shifting from fictional aural events to phonetic aural events, whether the rat-like rustle of *s*- and *r*-sounds in 'squeaking rapacious rats' or the shushed sibilance of 'slow and soft but not stealthy'. Justin St. Clair hears the 'novelistic rebroadcast' of any 'aural media' as 'almost constitutionally schizophonic'. If schizophonia 'is the splitting of sound from source', its 'linguistic retransmission' is a 'copy of a copy'. Rebroadcast sound in fiction thus splits the 'schizophonic reproduction [. . .] from *its* source', making a 'doubly refracted aural moment'.[26] Garrett Stewart's theory of phonemic reading extends this telescopic schizophonic effect.

Stewart asks 'where do we read?' The question seems insufficient, and he zones in: 'Where, then, does *where* leave us?' It leaves us, it transpires, in 'none other than the reading body', the 'somatic locus of soundless reception' which includes 'the brain but must be said to encompass as well the organs of vocal production, from diaphragm up through throat to tongue and palate'.[27] Where *where* leaves us, and where silent reading locates itself, is, then, 'in the conjoint cerebral activity and suppressed muscular action of a simultaneously summoned and silenced enunciation'.[28] Stewart locates his own argument in staunch opposition to the 'phonophobia' left in 'the wake of the Derridean attack on the Logos'.[29] If poststructuralism staged the death of the author, Stewart records some collateral damage:

> [Since we are no] longer disposed to hear the author behind the text, the Voice behind all inscription, we are at the same time (though not necessarily by the same logic) no longer encouraged to hear the phonemic counterpart to the spaced lettering of a text.[30]

To correct this kind of critical inattentiveness, Stewart advances 'phonemic reading', a practice 'to do not with reading orally but with aural reading'.[31] Phonemic reading pays attention to how texts secrete meaning through the sound-effects of syllables and sentences; it stages an aural performance of written text in readers' heads. Stewart dubs this dubbing the '"inner" articulation [. . .] or "endophony"' of a text and argues that 'only when the ear is activated by association with the graphic can any text, however arial, however ethereal, be said to go into production'.[32] Phonemic reading pricks the reader's ear toward a threshold between printed words on a page and immanently voiced sounds in the skull. It 'voices rhyme', whether 'the traditional modes of end [. . .] and internal rhyme' or the 'imperfect but no less rich rhyme [. . .] between inscribed and vocalised signifiers'.[33]

Rhyme effects, in which a sound plays, peters out, then plays again, produce their own schizophonic affect. The text transmits a sound then retransmits an iteration. Jacques Derrida calls rhyme 'the general law of textual effects [. . .] the folding together of an identity and difference'.[34] We will hear how *Infinite Jest* manipulates schizophonic rhyme effects across its storyworld and sentences in order to dramatise the phenomenology of this kind of folding-together – between self and other, identity and difference – in representations of schizoid and schizophrenic lifeworlds. It is essential to stress that this folding-together is a perceptual experience. As Paul and John Lysaker write, it is wrong to 'regard schizophrenia as an utter collapse of self'; schizophrenic people may 'find themselves threatened by and occasionally overwhelmed by intersubjective and powerful subjective experiences', but they 'maintain a sense of self'. The illness 'retains an irreducible, first-person dimension', which any 'adequate conception of the disorder' must track.[35]

Postmodern cultural theory has sometimes lost track of this irreducible, first-person dimension. As Angela Woods writes, it 'has used the term schizophrenia to ends far removed from the clinical diagnosis of individual suffering'.[36] This kind of misuse has consequences. In 2013, neuroscientist Patrick House observed that 'something strange is happening with the word *schizophrenia*'.[37] Illness can warp into a way of describing or denoting something separate from illness. Words 'drift from the language of disease into the language of everyday things [. . .] through metaphor'. For House, schizophrenia has become 'unlike any other word in the English language', to the extent that anything with 'contradictory or antagonistic qualities or attitudes' might be diagnosed as 'schizophrenic', whether a 'rapper [. . .] headlining as a poet', a volatile 'stock market', or even 'inclement' weather. He reels off stark data on schizophrenia's sticky prevalence in popular media, noting a 2003 study which claims that 'U.S. newspapers use *schizophrenic* metaphorically

28 percent of the time', and calculating that, between 2012 and 2013, *The New York Times* had meant the word metaphorically thirty-eight percent of the time. Even *Slate* magazine, which published his own article, had used the word metaphorically 'three dozen times since 1996'. House is right about all of this, but the present tense of his first quoted sentence seems wrong. By 2013, this 'strange' thing *had been* happening for decades. As Woods writes, schizophrenia was already very much on 'the agenda of cultural theory' in the 1970s and '80s, in the combined wake of both 'a strong counter-cultural anti-psychiatry movement' and the publication of Gilles Deleuze and Félix Guattari's *Anti-Oedipus: Capitalism and Schizophrenia* (1972).[38]

A crystallising, crystal-ball example of this took place over a long weekend in November 1975, when Sylvère Lotringer hosted the Schizo-Culture conference in New York. The conference's original aims were admirable. It sought to reclaim the term 'schizo', redeploying it in anti-psychiatric contexts to articulate ways of combatting controlling structures of psychiatric practice and capitalistic oppression. But, as Susan Sontag wrote two years after the conference, 'illness is not a metaphor, and [. . .] the most truthful way of regarding illness [. . .] is one most purified of, most resistant to, metaphoric thinking'.[39] The conference represents an influential and prophetic nexus of schizophrenia's metaphorical misuse. It was a big deal. Rockstar stardust dusted it, as varied worlds converged in raucous style. Political radicals, philosophers, ex-cons, rock bands, critical theorists, composers, psychiatrists, Black Panthers, and theatre troupes all jostled to be heard in an event that New York newspapers *The Village Voice* and *The Soho Weekly News* promoted like a boxing match. The former styled the event as an inevitable 'slugfest' and interviewed co-organiser John Rachjman, who promised 'a lot of shouting'.[40] A similar noisiness makes the conference's title hard to make out, and 'schizo' seems easy to mishear. *Anti-Oedipus* had been published in 1972, but it was not translated into English until 1977. Its notorious model of the 'schizo' is an idealised figure, a 'desiring-machine' that continually breaks down in a world where 'desiring-machines work only when they break down'. This schizo exists in opposition to the 'artificial schizophrenic found in mental institutions [. . .] produced as an entirely separate and independent entity'.[41] The conference accepts this fissure as geological fact. The problem, as Woods writes, is that 'schizo' and 'schizophrenia' are 'too closely connected, too intimately interlinked to be meaningfully separated'.[42]

Schizophrenia is the name of a mental illness, and 'schizo' is (or sounds like) slangy shorthand for that illness. The conference tries to magic these words apart. Its press release claims that 'schizo' does not refer 'to any clinical entity, but to the process by which social controls of all kinds [. . .]

are broken up and opened to revolutionary change'.[43] William Burroughs – conference attendee and speaker – backs it up:

> I think 'schizo-culture' here is being used rather in a special sense. Not referring to clinical schizophrenia, but to the fact that the culture is divided up into all sorts of classes and groups, etc., and that [. . .] old lines are breaking down.[44]

But 'schizo' does not function independently from 'schizophrenia'. Rather, it not only invites but insists upon vivid, visceral associations with the illness, throwing its voice like a ventriloquist. Their source is invisible, but still those three snipped syllables – *phre-ni-a* – catch stickily in the ear, profoundly and problematically audible. German physician Eugen Bleuler coined 'schizophrenia' in 1893. As we've seen, 'schiz' means 'split', while 'phrenia' traces back to the ancient Greek *φρήν* [phren], which referred originally to the diaphragm but spans 'breath', 'mind', head', 'soul', and 'spirit'. 'Schizo-Culture' involves a special type of surgery, then, in that cutting 'phrenia' from 'schizo' removes the body, brain, and soul from 'schizophrenia'. Without its thinking, breathing part, it becomes an illness without a human host. This bodiless form of schizophrenia is prismatic, plasticky; madness becomes malleable, a means for metaphor-making that depends upon a prissy, anempathetic pivot away from human brains and bodies. This bodiless illness – schizophrenia as aesthetic abstraction – infects some of postmodern critical theory's most influential texts.

The conference programme from 1975 gives a snapshot of how 'schizo' has pinballed jarringly at times between psychiatric and non-psychiatric contexts. It promises a workshop called 'Schizo-city (Harlem)'; a talk on 'The Schizo and the Institution' by psychiatrist François Peraldi; and, at the end, a 'Schizo-party'.[45] On one page, 'schizo' describes a district, a patient, and an after-party. Two years later, Lotringer edited a 'Schizo-Culture' edition of the journal *Semiotext(e)*, a 'modified version' of the event.[46] In its pages, schizophrenia spans disciplines in such a way as to mean pretty much anything. Peraldi's article describes the illness as a 'revolutionary process'[47]; for musician Boris Policeband, it's a 'solution' to the body's fascism, since 'it allows you to jump back and forth [. . .] without any sense of self'[48]; meanwhile, philosopher Alphonso F. Lingis describes his notion of '[s]chizophrenicized subjectivity' as a 'split, fragmented, dismembered, disintegrated field of momentary subjectivities, forming in pleasure and pain'.[49] These analogies elide schizophrenia's phenomenology. For as Louis Sass writes, schizophrenia 'may be characterised less by fusion, spontaneity, and liberation of desire than by separation, restraint, and an exaggerated cerebralism and propensity for introspection'. The 'schizophrenic dissolution of self is not the result of the disintegration of the ego; rather it is caused by an unseating and acute self-awareness'.[50] In other words, schizophrenia frequently involves a surfeit of self-reflexive thinking rather than its liberating, revolutionary absence. Compare this to Deleuze's description of schizophrenia as 'an involuntary and stupefying experience, something very, very acute, very intense, with high levels of intensity'.[51] Schizophrenia is involuntary, certainly, and intense often. But it isn't stupefying. There are repeating paranoias of stupefaction, self-automation, and inanimateness throughout *Infinite Jest*, and multiple characters experience themselves as empty, either unthinking or unfeeling. Burn quotes Laing to describe a condition 'near universal' in the novel, a disembodied way-of-being that is common to Laing's schizoid and schizophrenic person, in which people 'experience themselves as automata, as robots, as bits of machinery, or even as animals'.[52] The self-reflexivity of Laing's language is significant: a *way* of thinking or feeling is not a stupefied *absence* of thinking or feeling.

Erdedy is a comparatively minor character, whose thoughts are relayed through intermittent indirect free speech for ten pages near the book's beginning. He is well off, well educated, and morbidly addicted to marijuana. When we meet him, he has isolated himself in his flat. He has bought snacks. He is waiting for a woman who has promised to sell him two hundred grams of 'unusually good marijuana' with which he hopes to embark on a binge so extreme and unpleasant it will put him off the drug for life, a process he has attempted 'maybe 70 or 80 times before'.[53] He sits waiting in his apartment, his thoughts becoming more and more anaphorically compulsive, until the section ends with the sudden, simultaneous sounding of his telephone and front door's buzzer's intercom:

> because at this precise time his telephone and his intercom to the front door's buzzer both sounded at the same time, both loud and tortured and so abrupt they sounded yanked through a very small hole into the great balloon of colored silence he sat in, waiting, and he moved first toward the telephone console, then over toward his intercom module, then convulsively back toward the sounding phone, and then tried somehow to move toward both at once, finally, so that he stood splay-legged, arms wildly out as if something's been flung, splayed, entombed between the two sounds, without a thought in his head.[54]

The passage is so loudly and exaggeratedly *two-ish*. 'Time' chimes twice in the opening clause; 'both' sounds twice over successive clauses; we hear two pairs of 'towards' (two pairs of 'two-words'), and two variations of 'splay'. Then two more twos echo out of 'en*tomb*ed between the *two*'. The sentence plays for the reader the same sensory experience of sound Erdedy is subjected to in the storyworld, whether read aloud or not. As Stewart writes, although

> audible phonetic sounds do not result from a text [. . .] not read aloud, the 'inner' articulation – or 'endophony' – involved in silent reading [. . .] actuates the whole range of phonemic differentials' and 'latently engages' the equivalent 'somatic or muscular activity.[55]

The text's virtual soundscape interacts with its verbal soundscape, as the reader hears pairs of sounds recreate Erdedy's entombment between a pair of sounds. Film theorist Michel Chion delineates two broad types of music or noise in film: 'empathetic' and 'anempathetic'. Empathetic music participates 'in the scene, by taking on the scene's rhythm, tone, and phrasing', while 'anempathetic music' exhibits 'conspicuous indifference [. . .] progressing in a steady, undaunted, and ineluctable manner'.[56] Chion is talking about the relationship between an on-screen image and its soundtrack, but a similar effect occurs when the aural dimensions of a text syncopate with its storyworld. Wallace's sentence's soundscape is empathetic. It makes an echo chamber of entombing in-twos, which evokes Erdedy's aural entombment within the reader's inner ear.

This aural evocation of splayed indecision as self-division sets up one of *Infinite Jest*'s most prevalent concerns: the division of self associated with Laing's phenomenology of the schizoid personality. *The Divided Self* tracks the 'comprehensible transition from the sane schizoid way of being-in-the-world to a psychotic way of being-in-the-world'. By 'schizoid' Laing means 'an individual the totality of whose experience is split', for whom there is 'a rent in his relation with his world' and 'a disruption of his relation to himself'.[57] This is *not* a division into multiple personalities but rather a sense of rupture caused primarily by self-consciousness. Laing's schizoid personality embodies a certain ontological fragility, a susceptibility to feeling overwhelmed and impinged upon by the world, and Laing considered schizoid symptoms more like strategies: compulsive but comprehensible patterns of speech or thought that try to keep a sense of self intact. The ringing sounds Erdedy experiences in *Infinite Jest* condense and weaponise the three forms of anxiety that Laing identifies in the ontologically insecure person: 'engulfment, implosion, petrification'.[58] Engulfment refers to the dread of 'relatedness', in which a person fears they will lose their 'autonomy and identity' by opening themselves up to others; implosion is the perceptual experience that the world is 'liable [. . .] to crash in and obliterate all identity'; while petrification is the terror that one will be turned 'from a live person into a dead thing'.[59] Erdedy suffers each anxiety in rapid succession. The efforts of two people to communicate with him invade and empty his mind, turning him to stone. 'Entombed' envelopes each anxiety: it means literal surroundedness, contains a punningly audible 'into' suggestive of implosion, and summons the rocky stillness of petrification. The wordplay houses its own little ontological crisis, if we take Jonathan Culler at his word, for puns present 'the disquieting spectacle of a functioning of language where boundaries – between sounds, between sound and letter, between meanings – count for less [. . .] and where supposedly discrete meanings threaten to sink into fluid subterranean signifieds'.[60] 'Entombed' enacts this semantic sinking, and Erdedy is engulfed, sentenced to the same splayed petrification his name suggests: a hesitating 'er' that ends up 'ded'.

Erdedy's emptied-out state maps onto a prevalent paranoia in *Infinite Jest*. Characters fear or experience feelings of emptiness and petrification. Orin Incandenza considers himself 'an empty withered psychic husk', while Hal Incandenza hasn't had 'a bona fide intensity-of-interior-life-type emotion since he was tiny'.[61] Inside him 'there's pretty much nothing at all, he knows'.[62] But it is significant that the text gets louder as Erdedy's petrification looms larger. Erdedy's suffering sounds like ringing in the reader's ears, who is thus invited to participate empathetically with the scene and so becomes implicated in the undoing of Erdedy's petrification. Through our own participation in and co-production of the text's soundscapes, the reader's brain becomes the site of an intersubjective sound show. Sound effects become sounds that affect, showing Erdedy's perceptual experience of numb petrification to be just that: a perceptual experience. Laing writes that in 'the schizoid condition [. . .] there is a persistent scission between the self and the body'. The body can feel like 'the core of a false self, which a detached, disembodied "inner", "true" self looks on at [. . .]'.[63] Erdedy experiences disembodiment: he thinks of his bodily 'desires and ideas' and 'impulses being starved of expression [. . .] floating dryly away'.[64] But this disembodiment coincides with an increasing number of textual repetition and rhyme effects, which stress the sonic, sensory experience of reading, intensifying the sensation of voiced syllables sounding somewhere sort of in your skull. And since, as Stewart writes, even silent reading includes 'the brain [. . .] diaphragm [. . .] throat [. . .] tongue and palate', the bodily effect of such an auditory, affective scene represents a redemptive process of embodiment and empathetic intercorporeal identification for the reader.[65]

This aural, affective, and phenomenologically charged dramatisation of ontological anxiety in a schizoid personality contrasts vividly with representations of schizophrenia in influential critical theory. Fredric Jameson's 1983 'Postmodernism and Consumer Society' claims the figure of the schizophrenic as the arch-representative of postmodernity's phenomenology. As Woods notes, it had been cited over 7,560 times by 2011.[66] Its influence is enormous, yet its model of schizophrenia is scraggily unformed, constituting a kind of intellectual combover. Jameson's vision does not tally with clinical or firsthand accounts of the illness but evokes senseless, selfless fracture, echoing the Schizo-Culture conference by stressing schizophrenia's first two syllables to breaking point. Jameson announces his turn to the term with a disclaimer: 'I'm not even sure that the view of schizophrenia I'm about to outline – a view largely developed in the work of [. . .] Jacques Lacan – is clinically accurate; but that doesn't matter either, for my purposes'.[67] Jameson invokes Lacan's conception of schizophrenia as 'essentially a language disorder' and begins his argument somewhere in this realm, within a structuralist paradigm of sign, signifier, and referent.[68] Schizophrenia is 'the breakdown of the relationship between signifiers', so schizophrenic experience is 'an experience of isolated, disconnected, discontinuous material signifiers'; such a signifier 'has lost its signified' and 'been transformed into an image'.[69] From a linguistic standpoint, this visual characterisation is oblivious to some of schizophrenic language's most significant phenomenological features and signifying practices. As Giovanni Stanghellini writes, 'schizophrenic people actually have a more fluid semantic network which allows for an *expansion of the horizons of meanings* evoked by any given object or [. . .] word'. Indeed, they have 'a hyper-tolerance of semantic complexity' and 'actively seek out this broadening of the semantic halo'.[70] This synaesthetic willingness to *hear* punning semantic shimmering in words or things, a way of listening that sets words glistening, contradicts Jameson's rubble-like fragments, a deadened aftermath of failed signification.

Indeed, schizophrenic discourse frequently relies upon varied, complex modes of signification, often to do with aural qualities of language: rhythm, rhyme, puns, repetition, iconicity, or onomatopoeia. As Sass writes, schizophrenic people

> pay inordinate attention to the [. . .] sound of words [. . .] and they too are highly sensitive to puns [. . .]. They are also highly inclined to have a heightened awareness of ambiguities; to feel overwhelmed and confused by the multiplicity of potentials in language.[71]

Words can be splintery and unstable, their meanings liable to crack and shift punningly, as individual syllabic sounds pull seismically at the semantic flow of discourse. Elaine Chaika records a patient speaking about their mother:

> 'My mother's name was Bill [. . .] . . . and coo? St. Valentine's day is the official startin' of the breedin' season of the birds. All buzzards can coo. I like to see it pronounced buzzards rightly. They work hard. So do parakeets [. . .].'[72]

'Bill' sparks an association with birds, which sets off birdsong: '"coo?"' Coo has a double meaning, describing the trilling of pigeons or doves or the soft murmurs of Valentine's Day lovers. '"Buzzards"' descend, and the 'proper pronunciation' of the word's second syllable offers a rhyming possibility with 'hard', so: buzzards '"work hard"'. Syllabic sounds lead the way semantically, as 'poetic sparks' or 'idiosyncratic poetic associations [. . .] overwhelm normative semiotic interpretation'.[73] These poetic sparks can illuminate cryptic depths. Sass recalls one schizophrenic patient, asked about the death of their father. The patient switches from their father's 'demise' to their father's 'dim eyes'. As Sass writes,

> [to] shift from *demise* to *dim eyes* in writing about one's father's death is not to make a merely random sound association but to suggest something thematically sensible and rich – and this [. . .] is very often the case with schizophrenic glossomania.[74]

This elision of the complex ways words and syllables can interact to produce meaning extends to Jameson's more absolute elision of schizophrenic selfhood. Because, '[f]or Lacan, the experience of temporality, human time, past, present, memory, the persistence of personal identity over months and years – this existential or experiential feeling of time itself – is also an effect of language', the breakdown Jameson sees between signifiers breaks out from the safe confines of sentences. Language, too, 'moves in time,' and that syntactical temporality gives us 'a concrete or lived experience of time'. But 'since the schizophrenic does not know language articulation in that way', they are 'condemned to live in a perpetual present'.[75] The schizophrenic, then, 'is not only "no one" in the sense of having no personal identity; he or she also does nothing, since to have a project means to be able to commit oneself to a certain continuity over time'.[76] We might compare this literal lack of selfhood to the perceptual experience of this lack of selfhood in Julie, a schizophrenic patient of Laing's. She says, 'I'm thousands [. . .] I'm a no un', which Laing parses as 'nun: a noun: no one single person'.[77] An expression of perceptual emptiness is in fact dense with punning, signifying, and self-reflexive meaning.

Woods wonders what has to be 'downplayed, side-stepped, or even repressed' for Jameson's metaphor-making to function.[78] The simplest answer is selfhood: Jameson's schizophrenic has none. His argument relies on an extract from Renée Sechehaye's 1951 *Autobiography of a Schizophrenic Girl*. Swiss psychotherapist Marguerite A. Sechehaye (who became the author's adopted mother) contributes a psychiatric interpretation and commentary. Jameson disavows schizophrenia's clinical, psychiatric contexts, then pivots his argument on a passage from a first-hand account of schizophrenia, framed by a psychotherapist. The claims

Jameson makes of schizophrenic people – that they have 'an undifferentiated vision of the world in the present', and that they do 'not know [. . .] the persistence of the "I" and the "me" over time' – fit his own rather forgetful text more aptly.[79] The extract he quotes refutes his argument with its first letter: 'I remember very well the day it happened'.[80] The quotation is in the first-person singular, which it shows quickly to be retrospective and self-reflexive, not merely exhibiting but examining the persistence of its 'I' over time. It dispels the myth of depthless, atemporal selfhood Jameson proposes as characteristic of schizophrenia and postmodernity. This is just one sentence, and the spectre of the rest of Sechehaye's text hovers, eyebrows raised, over Jameson's essay. *Autobiography of a Schizophrenic Girl* is a coherent narrative that describes in rich detail the delusions, hallucinations, and sensory phenomena Renée experiences. The content of Renée's delusions is frequently incomprehensible, but that incomprehensibility is the point. The narrative navigates between delusional and non-delusional belief systems and, in doing so, demonstrates the personal identity of its author and protagonist and the persistence of this identity over time. Like Erdedy's splayed stupefaction between two sounds in *Infinite Jest*, the text presents the perceptual experience of ontologically obliterating engulfment. Renée cries out that 'water [. . .] was rising, rising higher, ready to engulf me'. But an irreducible first-person dimension remains, and the memoir ends in passionate self-expression: only 'those who have lost reality [. . .] can truly taste the joy in living and prize the transcendent significance of being a part of humanity'.[81]

Infinite Jest's next entombment involves Orin Incandenza, who finds the 'schizoid bounces' of American footballs inexplicably 'disorientating and upsetting'.[82] Erdedy ended his section 'without a thought in his head'. Likewise, Orin watches a dead bird bob in his pool 'without a conscious thought in his head'.[83] 'Entombed' echoes again, two times: he wakes one morning 'entombed in [. . .] psychic darkness'; the next, 'curled and entombed'.[84] Aural and physical scissions evoke divided selves around him. He leaves a petrified impression of himself 'sweated darkly into the bed', a 'white salty [. . .] fetal-shaped fossilised image'.[85] His body also divides in two, with an 'over-sized left arm and big left leg'.[86] He wakes another morning after a night of 'horror-show dreams [. . .] his wet silhouette on the bottom sheet like a coroner's chalk outline'.[87] Aural doppelgängers soundtrack this visual doppelgänger, with the claggy rhyme of 'silhouette' and 'wet' coupling with the alliterative consonance of 'coroner's chalk'. Orin has woken up next to a 'Subject', his name for the women he compulsively seduces. She's watching 'something horrific called "INTERLACE EDUCATIONAL CARTRIDGES IN CONJUNCTION WITH CBC EDUCATIONAL PROGRAMMING MATRIC PRESENTS *SCHIZOPHRENIA: MIND OR BODY?*"'.[88] Orin watches, 'paralyzed', as a cruelly colloquial voiceover introduces a patient, who is subjected to a sadistic experiment:

> And so but since the old CBC documentary's thesis was turning out pretty clearly to be SCHIZOPHRENIA: BODY, the voiceover evinced great clipped good cheer as it explained that well, yes, poor old Fenton here was more or less hopeless as an extra-institutional functioning unit, but that, on the up-side, science could at least give his existence some sort of meaning by studying him very carefully to help learn how schizophrenia manifested itself in the human body's brain.[89]

Again, pairs of phonetically similar sounds enact aural divisions within the sentence. 'CBC' immediately echoes the weird wastefulness of 'so but since', and we get the intertwined alliterative repetition of 'great clipped good cheer'. A 'brisk Albertan voice-over' speaks mercilessly over Fenton's suffering, explaining that he's 'a dyed-in-the-wool paranoid schizophrenic who believed that radioactive fluids were invading his skull'.[90] The experiment will 'advance science' by injecting him with 'special blood-brain-barrier-penetrating radioactive dye' and sticking him in a body-sized scanner. We then hear 'the machine's blurbs and tweets not even coming close to covering Fenton's entombed howls as his worst delusional fears come true in digital stereo'.[91]

Once more, the text inflicts each of Laing's ontological anxieties upon a suffering, divided subject. Fenton's howls are 'entombed' and in-two, mediated by the binary signal of 'digital stereo'. He is engulfed by the 'body-sized receptacle' and imploded by the radioactive dye, while the PET scanner's acronym ominously implies his petrification. The virtual soundscape here is anempathetic and schizophonic. The voiceover's slangy cheeriness speaks with sinister, disembodied indifference over Fenton's suffering, accompanied by the robotic bleeps and squeaks of the machine. Its verbal soundscape, however, is empathetic. Pathos plays audibly, as again the sound-effects of sentences syncopate with the storyworld to represent a specific (and loudly Laingian) type of ontological crisis. Sound breaches the scene's split forms of petrified spectatorship, in which a paralysed Orin watches a paralysed Fenton, precisely when it breaches the material confines of black-and-white text. Silent page becomes sounding board, as rhyme effects reproduce the storyworld's auditory entombment within the reader's head.

Like Jameson's essay, Jean Baudrillard's 'The Ecstasy of Communication' (1983) makes a silent, visual metaphor from schizophrenia, claiming the illness as analogy for the phenomenology of postmodernity. Rolling past metaphors like roadside pile-ups – from 'hysteria' as an expression 'of the body's theatrical and operatic conversion' to 'paranoia' as 'the pathology of organisation' – Baudrillard pulls up at schizophrenia. He equates the experience of the 'schizo [. . .] bereft of every scene' to the 'immanent promiscuity' of 'communication and information' that characterises the empty excess of modernity's material signifiers. With 'all these networks, with their continual connections, we are now in a new form of schizophrenia'.[92] Visibility consistently forms the frame of reference. Baudrillard's new obscenity – his state of schizophrenia – is 'the obscenity of the visible, of the all-too-visible, of the more-visible-than-the-visible'.[93] The symbolic realm of his argument is that of a light show: a photogenic, if morgueish, auditorium of mirrors and screens in which information is rendered as rays of light. Vision and consciousness converge along the same optic nerve, and the 'schizo' experiences 'the overexposure and transparence of a world which traverses him without obstacle'.[94] He can 'no longer produce the limits of his own being' but becomes 'a pure screen, a switching center for all the networks of influence'.[95] Both essays – Jameson's and Baudrillard's – describe a state of 'total instantaneity'.[96] But Baudrillard's schizo is effaced by the same external phenomena which Jameson's schizo is split from, as these powerfully influential essays (originally published side-by-side in Hal Foster's 1983 *Anti-Aesthetic: Essays on Postmodern Culture*) posit contradictory phenomenologies of schizophrenia. Yet their fictional schizos still share a likeness. Both are passive, locked-in lookers: pinned-open pairs of eyes that see ceaselessly, looking either into pitch-black or bright light, as the varied symptomology of schizophrenia leads to a high theory hijacking, a kind of critical theory car-boot sale, in which individual bits of the illness are picked up, cursorily examined, and cheaply bought, to be upcycled into all-encompassing analogies for contemporaneity.

For Jameson and Baudrillard, 'postmodernism is characterised above all by a loss of depth'. Everything is surfaces and screens, although Baudrillard's schizophrenic screen does not remotely resemble any 'specifically psychotic symptom'.[97] The emphatic visuality of both essays makes this surface sheen as wide as a horizon. As Walter J. Ong writes, if you want to 'test the physical interior of an object as interior, no sense works so directly as sound'. The human eye, however, is 'adapted best to light diffusely reflected from surfaces'; it does 'not perceive an interior [. . .] as an interior'.[98] Likewise, Jameson's ear does not extend to the cryptic depths of signification in schizophrenic discourse, while Baudrillard's ear does not do much at all. Rather, material signifiers traverse his schizo-screen solely through the eyes, obliterating the 'I'. His 'new state of things' sees 'the forced extroversion of all interiority', so any potentiality of sound to plumb depths becomes pointless.[99] However, it is the essays' formulations of schizophrenia that are surface deep. Even the existential perception of emptiness, of a lack of self, paradoxically disproves itself. As Laing writes, the 'denial of being' is 'the basic defence [. . .] in every form of psychosis, 'a means of preserving being'.[100] Wallace inherits the postmodern, poststructuralist preoccupation with schizophrenia. His 'body of work is noticeably filled with references to internal division [. . .] and schizophrenia', and 'his fiction is full of schizophrenics'.[101] However, *Infinite Jest* develops the postmodern tendency to *see* schizophrenia as cultural metaphor, moving toward phenomenologically dense, aurally inflected representations of disembodiment, engulfment, and petrification. Its soundscapes attune the reader toward empathetic participation in moments of ontological crisis; in doing so, the soundscapes go some way to restoring 'phrenia', the living, breathing, and thinking part of the word, to 'schizo'.

There is one more eruption of two schizophonic and entombing sounds near the novel's end. Don Gately is a recovering drug addict who works as an in-house staffer at Ennet House, a halfway house for people in recovery, where 'schizophrenia is like the norm'.[102] He gets shot defending a resident from a group seeking retribution for their murdered dog. In hospital, in excruciating pain, Gately remembers the experience that made him go sober, an intense binge on pharmaceutical painkiller Dilaudid. His friend and fellow enforcer Gene Fackelmann has stolen money and drugs from a violent gangster. The two sit in an apartment, injecting prodigious quantities of opiates and glazedly suffering the increasingly urgent attempts of the gangsters to contact them:

> When the phone rang it was just a fact. The ringing was like an environment, not a signal. The fact of its ringing got more and more abstract. Whatever a ringing phone might signify was like totally overwhelmed by the overwhelming fact of its ringing [. . .]. The intercom's buzzer to the luxury apartment complex's glass doors downstairs sounded, overwhelming both of them with the fact of its sound. It buzzed and buzzed. They discussed wishing it would stop the way you discuss wishing it would stop raining.[103]

The buzzing and ringing sounds are doubly schizophonic. Already disembodied, they become split not just from their source but from what they signify. Gately knows Fackelmann is in trouble; the sounds ought to convey danger, incite action. But meaning dissolves in Dilaudid in a way that echoes how some schizophrenic people experience the materiality of voiced sound. Deleuze describes how Louis Wolfson, schizophrenic author of *Le schizo et les langues* (1970), suffers certain syllabic sounds like shards of glass, which sharpen precisely *when* they shed connotative meanings. These linguistic 'fragments merge with unbearable sonorous qualities', and they become 'singularly wounding' at the exact 'moment' they are 'stripped of [. . .] sense'. A 'pure language-affect is substituted for the effect of language'.[104] Similarly, the words' phonetic qualities here cleave so closely to what they denote that sign speaks over signified: words like 'ringing' and 'buzzed' do ring and buzz, but their significance is lost over the sounds of the signs. Again, the repetitive sound effects of sentences recreate the auditory engulfment of the storyworld, merging effect and affect. The chiasmus of 'ringing [. . .] overwhelmed [. . .] overwhelming [. . .] ringing' puts two overwhelmed bodies in the middle of two ringing sounds. 'Overwhelmed' is overwhelmed into 'overwhelming', as the past participle gives way to the ringing 'ing' ending. We get the precise repetitions of 'buzzed and buzzed' [. . .] sounded [. . .] sound', and of 'discussed wishing it would stop [. . .] discuss wishing it would stop'. These eerily echolalic effects earmark a scene that is itself an echo, as the dual ringing sounds call back toward Erdedy's auditory, disembodying entombment. Gately's own disembodiment is narcotic rather than schizoid, but still it fulfils each of Laing's ontological anxieties: he's aurally engulfed by the ringings, and then the physical safety of the apartment is imploded: a 'fist [. . .] came through the window, spraying double-pane glass'.[105] Fackelmann's eyes are sewn open, and Gately is injected and incapacitated with 'pharm-grade Sunshine'; he 'can't feel his legs' and feels 'less high than disembodied'.[106] The last thing he sees is a reflection of his 'big square pale head with its eyes closing'.[107] Like Erdedy, Orin, and Fenton, Gately falls into Laing's final ontological anxiety: petrification. He is turned from a 'live person into a dead thing [. . .] an it'.[108]

Paul Crichton revisited *The Divided Self* in 2007. He wonders, upon reflection, whether its 'real value' is 'ethical rather than scientific', but insists that the text remains 'fundamentally important' as a model for understanding and respecting psychiatric patients.[109] Laing's final image is sunken treasure, a quotation from a schizophrenic patient who believes 'there was something of great worth deeply lost or buried inside her'. If one could 'go deep into the depth of the dark earth one would discover "the bright gold" [. . .] "the pearl at the bottom of the sea"'.[110] *Infinite Jest* follows Laing down. It fulfils Angela Leighton's expectation of the literary text as 'a bottomless well of noises', sounding the phenomenological depths of schizoid and schizophrenic selfhood with aural set-pieces that animate surface-stricken representations of postmodern schizophrenia in critical theory.[111] Human bodies are restored to human illness, as the text's soundscapes evoke what Sontag once felt and heard in the writing of Antonin Artaud: 'a kind of physiological phenomenology [. . .] of the microstructure of mental pain'.[112] The reader's body's brain becomes complicit, staging its own somatic, intersubjective soundings. As an Alcoholics Anonymous veteran with 'geologic amounts of sober time' implores Gately, the thing to do is 'sit down at meetings and relax and take the cotton out of your ears and put it in your mouth and shut the fuck up and just listen [. . .] really *listen*'.[113]

Notes

1 David Foster Wallace, *Infinite Jest* (London: Abacus, 2014 [1996]), p. 839.
2 Ibid., pp. 490, 740.
3 Ibid., p. 789.
4 R. D. Laing, *The Divided Self*, 4th edition (London: Penguin, 1990 [1960]), p. 17.
5 Stephen J. Burn, '"Webs of Nerves Pulsing and Firing": *Infinite Jest* and the Science of Mind', in *A Companion to David Foster Wallace Studies*, ed. Marshall Boswell and Stephen J. Burn (New York: Palgrave Macmillan, 2013), pp. 59–86 (65).
6 Quoted in D. T. Max, 'Final Destination', *The New Yorker* (11 June 2007), 54–71.
7 Quoted in Stephen J. Burn, *David Foster Wallace's 'Infinite Jest': A Reader's Guide*, 2nd edition (London: Continuum, 2012), p. 14.
8 Tom LeClair, 'The Prodigious Fiction of Richard Powers, William Vollmann, and David Foster Wallace', *Critique: Studies in Contemporary Fiction*, 38 (1996), 12–37 (34).
9 Burn, *A Reader's Guide*, p. 15.
10 Mary K. Holland, 'Mediated Immediacy in *Brief Interviews with Hideous Men*', in *A Companion to David Foster Wallace Studies*, pp. 107–30 (130).
11 Mary Shapiro, 'The Poetic Language of David Foster Wallace', *Critique: Studies in Contemporary Fiction*, 60.1 (2019), 24–33 (24, 25).
12 Ibid.
13 Ibid., p. 30.
14 Steven Connor, 'Feel the Noise: Excess, Affect, and the Acoustic', in *Emotion in Postmodernism*, ed. Gerhard Hoffmann and Alfred Hornung (Heidelberg: Universitätsverlag C. Winter, 1997), pp. 147–62 (147).
15 Wallace, *Infinite Jest*, p. 73.
16 Ibid., p. 72.
17 Ruth Salvaggio, *The Sounds of Feminist Theory* (New York: State University of New York Press, 1999), p. 22.
18 Wallace, *Infinite Jest*, p. 183.
19 Ibid., p. 189.
20 R. Murray Schafer, *The Soundscape: Our Sonic Environment and the Tuning of the World* (Rochester: Destiny Books, 1994), p. 88.
21 Schafer, *The Soundscape*, p. 90.
22 Paul Crichton, 'A Profound Duplicity of Life: Uses and Misuses of "Schizophrenia" in Popular Culture and Professional Diagnosis', *Times Literary Supplement* (31 March 2001), 14.
23 Wallace, *Infinite Jest*, p. 1057.
24 Ibid., p. 490.
25 Ibid., p. 491.
26 Justin St. Clair, *Sound and Aural Media in Postmodern Literature: Novel Listening* (New York: Routledge, 2013), p. 84.
27 Garrett Stewart, *Reading Voices: Literature and the Phonotext* (Berkeley: University of California Press, 1991), p. 1.

28 Ibid.
29 Ibid., p. 3.
30 Ibid.
31 Ibid., p. 2.
32 Ibid., p. 292.
33 Ibid., p. 99.
34 Jacques Derrida, *Dissemination*, trans. Barbara Johnson (Chicago: University of Chicago Press, 1981 [1972]), p. 277.
35 Paul Lysaker and John Lysaker, *Schizophrenia and the Fate of the Self* (New York: Oxford University Press, 2008), pp. 33, 2.
36 Angela Woods, *The Sublime Object of Psychiatry: Schizophrenia in Clinical and Cultural Theory* (New York: Oxford University Press, 2011), p. 183.
37 Patrick House, 'Schizophrenic Is the New Retarded', *Slate* (22 January 2013): <https://slate.com/technology/2013/01/schizophrenia-definition-and-metaphor-schizophrenic-does-not-mean-multiple-personalities.html> [accessed 22 June 2020].
38 Woods, *The Sublime Object of Psychiatry*, p. 194.
39 Susan Sontag, *Illness as Metaphor* (New York: Farrar, Straus and Giroux, 1978), p. 7.
40 Howard Smith and Brian Van Der Horst, 'Schizo-Culture', *Village Voice* (10 November 1975), 26.
41 Félix Guattari and Gilles Deleuze, *Anti-Oedipus: Capitalism and Schizophrenia*, trans. Robert Hurley, Mark Seem and Helen R. Lane (Minneapolis: University of Minnesota Press, 2008 [1972]), pp. 8, 5.
42 Woods, *The Sublime Object of Psychiatry*, p. 160.
43 [Anon.], 'Press Release', repr. in *Schizo-Culture: The Event*, ed. Sylvère Lotringer and David Morris (South Pasadena: Semiotext(e), 2013), p. 7.
44 Quoted in Lotringer and Morris (ed.), *Schizo-Culture*, p. 161.
45 'Schizo-Culture 1975 Poster', *Monoskop*: <https://monoskop.org/images/3/39/Schizo_culture_1975_poster.jpg?> [accessed 22 June 2020].
46 Sylvère Lotringer, 'My '80s: Better Than Life', *Artforum*, 41.8 (2003): <www.artforum.com/print/200304/my-80s-better-than-life-4509> [accessed 22 June 2020].
47 François Peraldi, 'A Schizo and the Institution (a Non-Story)', *Semiotext(e)* [*Schizo-Culture* special issue], 3.2 (1978), 20–31 (28).
48 Boris Policeband, 'Antidisestablishment Totalitarism' (interview with Sylvére Lotringer), *Semiotext(e)* [*Schizo-Culture* special issue], 65.
49 Alphonso F. Lingis, 'Savages', *Semiotext(e)* [*Schizo-Culture* special issue], 96–107, 112.
50 Louis Sass, *Madness and Modernism: Insanity in the Light of Modern Art, Literature, and Thought* (Cambridge, MA: Harvard University Press, 1992), p. 10.
51 Gilles Deleuze in 'Capitalism and Schizophrenia', Félix Guattari, *Chaosophy*, ed. Sylvère Lotringer, trans. David L. Sweet, Jarred Becker and Taylor Adkins (South Pasadena: Semiotext(e), 2009), 53–68, 62.
52 Burn, '"Webs of Nerves Pulsing and Firing"', p. 73.
53 Wallace, *Infinite Jest*, p. 18.

54 Ibid., p. 27.
55 Stewart, *Reading Voices*, p. 7.
56 Michel Chion, *Audio-Vision: Sound on Screen*, trans. Claudia Gorbman (New York: Columbia University Press, 1994), 8.
57 Laing, *The Divided Self*, p. 17.
58 Ibid., p. 43.
59 Ibid., pp. 45, 46, 47.
60 Jonathan Culler, 'Introduction: The Call of the Phoneme', in *On Puns: The Foundation of Letters*, ed. Jonathan Culler (Oxford: Blackwell, 1988), pp. 1–17 (3).
61 Wallace, *Infinite Jest*, p. 288.
62 Ibid., p. 694.
63 Laing, *The Divided Self*, p. 69.
64 Wallace, *Infinite Jest*, p. 26.
65 Stewart, *Reading Voices*, p. 1.
66 Woods, *The Sublime Object of Psychiatry*, p. 3.
67 Fredric Jameson, 'Postmodernism and Consumer Society', in *Postmodern Culture*, ed. Hal Foster, 2nd edition (London: Pluto Press, 1985), p. 118.
68 Ibid.
69 Ibid., pp. 119, 120.
70 Giovanni Stanghellini, *Disembodied Spirits and Deaminated Bodies: The Psychopathology of Common Sense* (New York: Oxford University Press, 2004), p. 101.
71 Sass, *Madness and Modernism*, p. 201.
72 Elaine Chaika, 'A Linguist Looks at "Schizophrenic" Language', *Brain and Language*, 1 (1974) 257–76 (260).
73 James Goss, 'Poetics in Schizophrenic Language: Speech, Gesture and Biosemiotics', *Biosemiotics*, 4.3 (2011), 291–307 (294).
74 Sass, *Madness and Modernism*, p. 206.
75 Jameson, 'Postmodernism and Consumer Society', p. 119.
76 Ibid., pp. 119–20.
77 Laing, *The Divided Self*, p. 204.
78 Woods, *The Sublime Object of Psychiatry*, p. 186.
79 Jameson, 'Postmodernism and Consumer Society', p. 120.
80 Marguerite Sechehaye, *Autobiography of a Schizophrenic Girl*, trans. Grace Rubin-Rabson (London: Meridian, 1994), p. 21.
81 Ibid., pp. 84, 136.
82 Wallace, *Infinite Jest*, p. 289.
83 Ibid., p. 44.
84 Ibid., pp. 42, 44.
85 Ibid., p. 43.
86 Ibid.
87 Ibid., p. 47.
88 Ibid.
89 Ibid.
90 Ibid., p. 48.
91 Ibid.

92 Jean Baudrillard, 'The Ecstasy of Communication', trans. John Johnston, in *Postmodern Culture*, pp. 126–34 (132).
93 Ibid., p. 131.
94 Ibid., p. 130.
95 Ibid.
96 Ibid., p. 133.
97 Woods, *The Sublime Object of Psychiatry*, pp. 188, 191.
98 Walter J. Ong, *Orality and Literacy: The Technologizing of the Word* (London: Routledge, 1995 [1982]), p. 71.
99 Baudrillard, 'The Ecstasy of Communication', p. 132.
100 Laing, *The Divided Self*, pp. 149–50.
101 Burn, '"Webs of Nerves Pulsing and Firing"', 70–1.
102 Wallace, *Infinite Jest*, p. 435.
103 Ibid., p. 936.
104 Gilles Deleuze, *Logic of Sense*, trans. Constantin V. Boundas, Mark Lester and Charles J. Stivale (London: Bloomsbury, 2017), p. 90.
105 Wallace, *Infinite Jest*, p. 974.
106 Ibid., pp. 970, 980, 981.
107 Ibid., p. 981.
108 Laing, *The Divided Self*, p. 46.
109 Paul Crichton, 'Medical Classics: *The Divided Self*', *BMJ: British Medical Journal*, 334.7586 (2007), 211.
110 Laing, *The Divided Self*, p. 205.
111 Angela Leighton, *Hearing Things: The Work of Sound in Literature* (Cambridge: The Belknapp Press of Harvard University Press, 2018), p. 25.
112 Susan Sontag, 'Preface', in Antonin Artaud, *Selected Writings*, ed. Susan Sontag, trans. Helen Weaver (Berkeley: University of California Press, 1988), p. xxi.
113 Wallace, *Infinite Jest*, p. 353.

9 Teju Cole's 'Art of Listening'

Rachel Farebrother

Teju Cole's *Open City* (2011), a novel without a plot that comprises the digressive musings of Julius, a Nigerian-German immigrant who spends most of his time walking the streets of New York, has occupied a central position in discussions about new directions in Black US fiction in the 2010s. Tracking the emergence of what she terms 'The New Black Novel', Erica R. Edwards singles out Cole's mapping of the city to highlight a broader preoccupation with 'the spatial logics of the long war on terror'.[1] Another rich seam of analysis has explored *Open City*'s engagement with the promises and perils of cosmopolitanism. Madhu Krishnan, Werner Sollors, and Pieter Vermeulen, among others, have seized upon the tensions that animate Cole's thematic preoccupation with 'cosmopolitan curiosity', especially in 'open-ended conversations' between Julius and the people he meets (many of them immigrants) in New York and Brussels.[2] These thematic preoccupations cannot be disentangled from *Open City*'s collage form, which is, as Cole explains, 'completely infected with other readings, of texts, of films, of pieces of music, and other artifacts'.[3]

Cole's enduring fascination with inter-artistic forms, especially the interplay between fiction and photography, has garnered substantial critical attention.[4] In addition to his restless experimentation with diverse cultural forms and platforms – from Twitter, Spotify, and Instagram to the short story and the phototext – Cole is a photographer and an art historian who penned a regular column entitled 'On Photography' in *The New York Times* until February 2019. Not surprisingly, Cole's fiction bears the imprint of his preoccupation with visual culture in terms of both subject matter and form.[5] *Open City* is replete with allusions to paintings and photographs by the likes of Johannes Vermeer, Pieter Brueghel, and Martin Munkácsi. As Alexandra Kingston-Reese argues, the novel's 'stylistic and tonal variation' reproduces photographic effects such as 'slowness, shock, and shadowing'.[6] Indeed, in various interviews, Cole has spoken of 'bringing slowness into basically everything I do', an aesthetic project that, in fiction, entails the evocation of genres such as the still life, a focus on how

DOI: 10.4324/9780429291616-12

art is experienced or felt (through hearing and touch as well as sight), and a narrative mode that draws inspiration from the attentive, patient observation that he admires in the writing of art historians like Michael Baxandall and Svetlana Alpers.[7]

Finally, it is a mark of the diverse critical approaches provoked by this protean novel that Cole's refusal to supply a portable political message or to adhere to the generic conventions of 'the ethnic bildungsroman or the immigrant plot' has been presented as a turning point in critical debates about the Black diaspora.[8] *Open City* is a central plank in Yogita Goyal's insistence that 'new migrations demand new conceptualizations of diaspora' that depart from formulations of community grounded in racial solidarity or political collaboration.[9] She draws inspiration from a generation of African immigrant writers including Chris Abani, Chimamanda Ngozi Adichie, and Dinaw Mengestu who have challenged romanticised narratives of racial belonging to rethink race and racial formation in the age of globalisation and racial capitalism. An 'African novel' in which Nigeria is represented only in fragmented memories, *Open City* requires a fresh critical approach: 'the postcolonial state and the US racial state must be read together', Goyal argues, 'as must the histories of colonialism and slavery'.[10]

With the notable exception of Rachel Sykes's analysis of the 'quiet contemporary American novel' and Josh Epstein's illuminating exploration of *Open City* as a 'deconstructed *bildungsroman* preoccupied with Gustav Mahler's late compositions', critics have largely overlooked Cole's preoccupation with what Julius calls 'the art of listening'.[11] Not only is listening (or, more accurately, impaired listening) crucial to Julius's career as a psychiatry resident (and Cole's attendant fascination with the shifting dynamics between patient, therapist, and reader), but inadequate or distracted listening emerge as recurring tropes in the novel's exploration of art, identity, and crisscrossing cultures. In one memorable sequence, Julius encounters paintings by the Deaf artist John Brewster in the 'noiseless' space of the American Folk Art Museum, an experience that prompts him to wax lyrical on the 'silent transaction between artist and subject' and the 'quiet' expressiveness of what he calls 'artists of stillness'.[12]

Cole explores concepts of silence, stillness, and deafness by way of an engagement with visual art, a move that provokes vexed questions about the relationships between sound, image, memory, and ideology. *Open City* is structured around a series of tableaux or still images – or, to use Cole's term, 'choreography' – that enact productive slippages between word, image, and sound (or silence).[13] Analysis of the dynamic interplay between sound and silence, listening and deafness in *Open City* highlights complex entanglements between seeing and hearing that have been obscured because of a tendency to focus on the visual registers through which race is often configured in the United States. In a recent blog on the joy of reading phototexts, Cole discussed the 'quiet consolations of phototexts' that leave readers 'mesmerized by the silent and unanticipated intensity' in a 'world of deafening images', phrasing that underlines the readiness with which Cole turns to concepts related to sound and hearing in his analysis of visual culture.[14]

Such scrambling of the senses is part of a broader challenge to what Jonathan Sterne calls, in *The Audible Past*, the 'audiovisual litany', whereby the apparent battle between the senses is reduced to a 'zero-sum game' and 'differences between hearing and seeing are often considered as biological, psychological, and physical facts' rather than historically contingent concepts 'worthy of philosophical, historical, political, and ethical reflection'.[15] But the intersection between sound and image in Cole's fiction, critical essays, and phototexts is neither abstract nor philosophical. It is integral to a deep concern with the ethics of critical listening. As an artist and critic, Cole places a premium upon meditative attention to 'visual noise': in a 2016 essay on emergent concepts of heroism and agency in photography documenting the Black Lives Matter movement, he argues that it is the role of the cultural critic 'to listen to the noise of life without being deafened by it'.[16]

The ethical imperatives of critical listening also encompass readers and interpreters. Dynamic interplay between word, sound, and image demands readers' openness to the multiple, often contradictory meanings generated by an aesthetic of assemblage and juxtaposition. Such forms prompt readers to 'think constellationally', to borrow a phrase from one of Cole's most famous critical essays, a mode of attention that is underpinned by an ethical concern with interconnection and interdependence in the midst of division and conflict.[17] This chapter contends that analysis of 'the art of listening' in Cole's diverse oeuvre not only illuminates the significance of his formal choices, which tend towards heterogeneity, but it also lays bare a deep-rooted preoccupation with ethics that is best captured in the active interpretative role assigned to readers who are primed to make connections, spot allusions, and identify evasions and inconsistencies.

Julius's Listening

Julius 'learn[s] the art of listening' from his terminally ill friend and mentor, Professor Saito, a gay Japanese American expert in medieval literature who was interned at the Minidoka Camp in Idaho during the Second World War.[18] In probing conversations with Julius at his New York apartment, which is 'dense with its various collections' including 'South Seas masks', 'overstuffed bookshelves', and 'little figurines and puppets', Saito provides lessons on memory and history.[19] He takes a long historical view to illuminate contemporary issues such as 9/11 and the Iraq War while insisting that 'forgetting doesn't take long'.[20] Recalling the experience of 'committ[ing]' poems to memory during his internment, Saito teaches Julius 'the value of memory, and how to think of it as mental music, a setting of iambs and trochees'.[21]

Such statements attest to the potential for productive interplay between sound, word, and image when developing conceptual frameworks for interpreting the past, but they also anticipate Julius's privileging of aesthetic pleasure over ethical responsibility. Indeed, any conception of an idealised aesthetic realm is undercut by the colonial and capitalist assumptions that underpin Saito's 'lifetime of collecting'.[22] Julius also notices that photographs of family and friends are nowhere to be seen in the apartment, an observation that foreshadows the gaps and evasions in Julius's narration. Ironically, Julius's insightful interpretation of Saito's conversations and collections (and the counterpoint between them) establishes an interpretative template focused on the interplay between what is seen and heard that readers can apply to the narrative more generally.[23] The method of scrutinising tensions and correspondences between Saito's curation of art objects and memories puts into practice another lesson Julius has learned from his mentor, namely 'the ability to trace out a story from what was omitted'.[24]

In keeping with the multivocal form of *Open City*, Cole's portrayal of Julius as a listener is multidimensional. Julius's diagnosis and treatment of his psychiatric patients – which is based upon attentive listening to 'what the patient tells' him even though 'the mind is able to deceive itself' and 'the strongest symptoms are sometimes not visible' – is never depicted in the text.[25] Instead, it is refracted through philosophical reflections upon such matters as blindness, trauma, grief, sound, and the lives of individual patients. Although he 'indulge[s]' friends with 'humorous tales from the horror of mental illnesses', Julius's work as a psychiatrist during 'an epidemic of sorrow' is characterised by acute sensitivity to suffering.[26] Listening to what people say is integral to psychoanalysis, and Julius's recourse to what Peter Gay calls 'relatively anonymous and attentively passive' listening creates space for subtle exploration of 'doubts and questions' while resisting straightforward, causal narratives about identity and selfhood.[27]

Nevertheless, any idealisation of the curative powers of listening is complicated by the unstable relationships between analyst, patient, and reader in *Open City*. As Cole has explained, the stream-of-consciousness narration that is so characteristic of the novel takes its cue from the 'formalized testimony' of a patient on 'a psychiatrist's couch'. By placing a reader in the position of an analyst who witnesses Julius's 'oversharing and evasion' in a series of visits to his psychiatrist, the limitations of the talking cure become painfully apparent as Julius's 'self-deceiving self-awareness' is exposed.[28]

There is also a sustained focus on Julius's acute sensitivity to a variety of sounds and music styles, including a set-piece performance of Mahler's Ninth Symphony with Simon Rattle in Carnegie Hall, two *erhu* players on Broadway, and the interleaving of his voice with 'the murmur of the French, German, or Dutch radio announcers' when he reads aloud with European radio in the background.[29] Julius's responsiveness to cultural expression contrasts with the self-absorption and distraction he displays when in conversation. To take just one example, Julius listens to his mother's account of the painful circumstances of her early life in post-war Germany 'with only half an ear'.[30] His solipsism means that it does not dawn on him that his grandmother was raped by the Red Army when pregnant with his mother until he 'became interested in these things for [his] own sake', a carefully weighted phrase that captures a tendency to prize intellectual curiosity above all else.[31]

Given these associations between listening, empathy, and ethics, listening emerges, as Jennifer Stoever has argued, as 'a dynamic historical and cultural practice, an embodied critical sense shaping how and what we think, *and* an ethical act shaped by our thoughts, experiences, and ideologies, one both subject to discipline and offering agency'.[32] Consequently, it is important to reckon with another dimension of 'the art [or artifice] of listening': Julius's taste for masquerading as a 'sympathetic ear'.[33] When he visits an immigration facility run by Wackenhut with a church group called 'the Welcomers', for instance, he presents himself as 'the listener, the compassionate African who paid attention to the details of someone else's life and struggle', a 'story' concocted partly for the benefit of his on-off girlfriend, Nadège.[34] The performativity inherent to such listening comes into sharp focus when Julius describes assiduous responsiveness as a 'party trick'. Such triviality sits uneasily with awareness that Julius's interlocutor, Professor Saito, is dying: 'It was a kind of party trick, to continue a conversation of this kind and remain the whole while perfectly distracted. It was like a film in which the soundtrack and images were out of sync'.[35]

Earlier in the narrative, the juxtaposition of Julius's decision to 'shut the window' to muffle the sound of a Take Back the Night protest with Nadège's 'strained voice' on the telephone – 'in counterpoint with the protestors down below' – is part of a larger pattern of associations between sexual violence, female protest, and a refusal to listen.[36] In short, Cole's thematic and formal preoccupation with 'the art of listening' is entangled with ethical questions about responsibility and complicity. The full extent of Julius's detachment and evasiveness emerges only towards the end of the novel, when Moji, the sister of a school friend, confronts him about raping her when she was fifteen years old. As Cole explains in a 2013 interview, motifs of entrapment, enclosure, and sexual violence serve as

> quiet foreshadowings [. . .] that won't be caught on a first reading – they just register as part of the apparently aimless drift of the book – but that on a second reading will be as vivid as blood in a sink.[37]

To date, critics have tended to argue that Julius 'silences the many voices he invites to speak', as Rachel Sykes puts it.[38] Yet it is perhaps more accurate to say that he maintains tight control of the narrative by refusing to listen or by electing to remain silent, a pattern of behaviour established early in the novel. On a second reading, recurring motifs of windows, barriers, incarceration, 'sectioning off', and coercion loom large, prompting readers to reflect on their willingness to be lulled into a false sense of security by Julius's erudite, lyrical, and apparently artless reflections on the world.[39] When she confronts him, Moji tells Julius that he has 'been ever-present in her life, like a stain or a scar'.[40] Her voice modulates to 'a strained, shattered tone' that recalls earlier conversations with Nadège, creating threads of connection that provoke concern about his relationships with women.[41] With a callous disregard for Moji's distress, Julius offers no response to her accusation, opting for evasion. Eschewing the confessional mode that readers have come to expect, he serves up an arcane anecdote about Nietzsche and Camus, in which Moji's reference to pain and scarring is warped into a self-absorbed, trivialised form. In accordance with a broader preoccupation with the ethical responsibilities of the listener in *Open City*, this callousness towards individual suffering in the pursuit of cultural or intellectual enquiry ricochets back on the reader. As Rebecca Clark has noted, 'the ease with which critics have folded Julius's rape of Moji into the "narrative pattern" – a move that the novel doesn't exactly discourage – points to something even more central about the all-too-easily-incorporated act of interpersonal violation'.[42] Taken together with the exposure of Julius's propensity for bad listening, the thematic preoccupations with listening and conversation in the novel adumbrate an alternative, more ethical interpretative approach.

Word, Image, and Sound in *Open City*

Since readers are invited to weigh up the ethical implications and asymmetrical power dynamics of Julius's inadequate listening by making intratextual connections and comparisons, an understanding of Julius's bad listening fosters productive modes of attention a reader may be positively prompted to bring to the text, most notably by spotting allusions. As Cole has pointed out, *Open City* is best understood as a collage or collection:

'The events, sensations and thoughts are calculated to look as if they are random, but they are actually carefully selected. There's a curatorial instinct driving the writing'.[43] For one thing, the novel is packed with allusions to cultural theory, literature, philosophy, music, and visual art. Julius's narration includes astute commentary on everything from Walter Benjamin, J. M. Coetzee, Sigmund Freud, Tahar Ben Jelloun, and Primo Levi to Nietzsche, Edward Said, Mahler, Judith Weir, Michael Jackson, Yoruba cosmology, Johannes Vermeer, and Robert De Niro. Beyond Julius's erudite musings, it is difficult to miss Cole's reworkings of well-known literary texts such as the epiphany at the end of James Joyce's 'The Dead', which he transposes to rainy Brussels.

Readers can judge the scope and subtlety of Cole's aesthetic of assemblage by consulting an interactive version of the opening pages of *Open City* posted on the website *Poetry Genius*, which features 'annotations' verified by Cole. The notes identify, among other things, direct references to *Beowulf* and the composer Rodion Shchedrin, but they also bring to light less obvious debts to rhythms, imagery, sounds, and ideas in the writings of Ernest Hemingway, Julian Barnes, Kazuo Ishiguro, and the verse of the rapper Talib Kweli.[44] Cole's techniques of fragmentation, recontextualisation, and absorption assign readers an active interpretative role because, as Elza Adamowicz explains, disparate elements remain 'in a dynamic process of multiple meanings and hovering significations constantly reactivated'.[45] At least part of the text's meaning depends upon a reader's ability to identify and decode intertextual fragments so each element of the montage is interpreted in relation to other fragments, to the whole and to its origin (even if such interpretative practices are muddied by the fact that allusions are often 'densely packed, implicit rather than explicit', to borrow a phrase from Cole's account of his aesthetic practice in the 2018 phototext, *Blind Spot*).[46]

The centrality of intertextuality to Cole's project is underlined by the title 'op cit' – a Latin term meaning 'in the work cited' that also substitutes for *Open City* itself – that he gave to a Tumblr page completed in March 2011.[47] Inviting readers to reach beyond the confines of the text in their interpretations of the novel, the page features a beautifully curated collection of photographs, recordings, and written texts cited in the novel.[48] Cole added lyrical, suggestive titles to each item on the page. An 1818 address about the philosophical implications of sign language and intermedial forms of communication has the caption 'language formed of gestures', while a Sarah Vaughan recording that plays in the background when Moji accuses Julius of rape is accompanied by the ambiguous caption 'with strings', an allusion to the lush strings in Don Costa's arrangement of 'Spring Can Really Hang You Up the Most' (1962) and Julius's studied evasion of ethical responsibility.

The dynamic interplay between different cultural forms on the Tumblr page not only signals Cole's enduring fascination with inter-artistic forms and productive tensions between sound, speech, touch, and sight; it also encapsulates an aesthetic practice that unites Cole's work across diverse media including fiction, phototexts, Twitter, and curated Spotify lists, whereby fragments from one text are reconstituted and transposed into a new format. In interviews about *Blind Spot*, Cole uses the word 'bloom' to describe this technique, a term that recalls Walt Whitman's organic metaphors for dynamic poetic expression that outlives its author while underlining his enduring 'faith in the fragment and the montage'.[49]

More relevant to this discussion of the modes of critical attention invited by *Open City* is the dynamic interplay between sound and image in scenes that represent the experience of encountering works of art in a gallery, museum, or concert hall. Taking his cue from art historians like Michael Baxandall who are as much concerned with 'thought after seeing the picture' as with the image itself, Cole challenges what he sees as 'an unwritten rule against bringing in other cultural experiences into a book', including 'the experience of buying books, reading books, going to museums'.[50] Drawing inspiration from sixteenth-century Dutch art, the subject of his doctorate, Cole explains that a desire to represent the aesthetic and psychological dimensions of such encounters prompted him to privilege narrative stillness over plot: he sought 'to slow things down to the point where [he could] give an account of what's happening when a person is looking at something [. . .] not just aesthetically but psychologically, too'.[51]

The prominence of sound and impaired listening in these fictional explorations of the affective, philosophical, and sensory dimensions of viewing art is best captured when Julius visits art galleries and exhibitions. Early in the novel, Julius goes to the American Folk Museum, where he stumbles upon an exhibition of paintings by John Brewster, who was profoundly Deaf. Julius's imaginative responses to the paintings are presented as a kind of thinking and feeling with the ears. The paintings – portraits of 'self-possessed', 'doll-like' children – create 'an air of hermeticism', 'stillness', 'silence', and 'quiet'.[52] Julius's descriptions of specific portraits pay careful attention to visual elements such as colour and form. When he discusses *One Shoe Off* (1807), for instance, 'a painting of a child holding a bird on a blue thread', he notices a sharp contrast between the 'muted' colour palette and 'the electric blue of the thread', picking out details such as the 'carefully observed lace frill' to speculate that the picture might be interpreted as 'a talisman against death' at a time of high child mortality.[53] But sound (and deafness) is the main critical framework for interpreting these portraits. Indeed, Julius compares the romanticisation of blindness as 'unusual sensitivity' or 'genius', a trope he finds in British, American, and Nigerian culture, with narratives that cast deafness as 'unfortunate' or even 'pejorative'.[54] Julius bucks this trend by identifying a tradition of visual art defined by 'stillness' and 'quietness': 'A laden brush, in depositing paint on the panel or canvas, hardly registers a sound, and how great is the peace palpable in those great artists of stillness: Vermeer, Chardin, Hammershøi'.[55]

According to Julius, Brewster's painting is unique because it 'communicate[s] the silence of his world' without resorting to techniques such as 'chiaroscuro': 'his deafness made him an outsider, and his images were imbued with what that long silence had taught him: concentration, the suspension of time, an unobtrusive wit'.[56] The quality of stillness in the paintings bleeds into Julius's solitary experience in the gallery, where he is alone apart from a single guard. Not only do Brewster's paintings render his 'mind quiet', but he internalises them in ways that highlight Cole's recourse to sound and its absence when representing the aesthetic and psychological effects of viewing art.[57] After leaving the gallery, Julius is haunted by one particular painting. To portray the pleasures of music as 'a sealed-away world, visible from without, but impossible to enter', he evokes sensations associated with touch and sight, conjuring up music's absence through striking visual imagery to create a complex sense of intermediality:

> I thought of Brewster's portrait of the deaf teenager Sarah Prince at the pianoforte, an instrument that neither artist nor sitter would have heard: the quietest piano in the world. I imagined her running her hand along the keys but refusing to press down on them.[58]

Alongside scenes that represent Julius's encounters with visual art or music (such as his visits to an exhibition of the Hungarian photographer Martin Munkácsi at the International Photography Center and to a Mahler concert in Carnegie Hall), the limitations of Julius's perception of the people he encounters in his everyday life are explored through a comparable focus on resonances and mismatches between sound and image. It is, for example, easy to see direct correspondences between the patient, lyrical descriptions of Julius's visit to the American Folk Museum in New York, and his brief trip to the Notre Dame de la Chapelle in Brussels. In both scenes, Julius enters the building to shelter from the rain, paying close attention to light, architectural space, and colour, imagining himself to be 'alone'.[59] Julius's first impression on entering the church is of 'total silence', but he soon becomes aware of an organist playing Peter Maxwell Davies.[60] What he imagines to be an 'unseen organist' is in fact a recording 'piped in through tiny speakers'.[61] Patterns of misperception and misinterpretation continue as he detects a 'dissonance in the sound of the organ music', 'distinct fugitive notes that shot through the texture'.[62] Eventually, he identifies the 'source of the fracture in the sound: a small yellow vacuum cleaner' operated by a woman who may be Rwandan or Congolese.[63]

Such painstaking tracking of Julius's perception captures in its dynamics the disruption of a romantic view of 'silence' as a retreat from the 'incessant loudness' of modernity.[64] Julius speculates about the woman's past, wondering if she 'might be here in Belgium as an act of forgetting' or seeking 'a refuge' from trauma, but her interior life remains unreadable and opaque.[65] In a series of details that invite readers to make connections with Julius's earlier reflections on Brewster's paintings, Julius picks out items of clothing (such as the unnamed woman's 'bright green scarf'), focusing on surfaces, fabrics, and colours to create a striking visual image.[66] He invokes the tradition of 'artists of stillness', connecting the woman cleaning the church with 'those women that Vermeer painted in this same gray, lowland light; like theirs, her silence seemed absolute'.[67]

To some extent, the refusal of legibility that animates this picture in words might be read as a mode of resistance. It recalls Cole's praise, in 'A True Picture of Black Skin' (2015), of Roy DeCarava's emphasis upon the 'opaque, dark, shadowed, obscure' in his photography, aesthetic choices that eschew demands for exposure and hypervisibility to 'make a case for how indirect images guarantee our sense of the human'.[68] In that essay, Cole mounts an argument for the right to remain obscure by drawing upon Édouard Glissant's theorisation of 'opacity', a concept he parses as 'a right to be misunderstood if need be'.[69]

The soundscape of the Brussels church, which is characterised by 'fracture', 'dissonance', and 'fractured, scattered feeling[s]', pulls in another direction, unsettling Julius's incorporation of the woman's image into an aesthetic of 'stillness'.[70] Jarring notes and emotions limn the limitations of any reading that places a premium upon what King-Kok Cheung has called 'articulate silences'.[71] For one thing, Julius displays no interest in speaking with the woman, partly because emotional and critical distance allows him to aestheticise suffering without attending to the messiness of individual lives or emotions. There are also revealing affiliations and substitutions between the display and viewing of cultural artefacts and people in Julius's beautifully composed word picture. Finally, the 'wavering hum' of the vacuum cleaner highlights the woman's work as a cleaner, shifting attention towards how beauty is sustained by various kinds of labour, whether that be a cleaner in the church or the guard Julius assiduously ignored during his earlier visit to the American Folk Museum.[72]

In *Open City*, the interplay between word, image, and sound generates aesthetic and ethical dissonance (and unease) that encourages readers to practise a kind of critical listening. Cole's intermedial forms, which combine textual allusions, visual effects, silence, resonance, and dissonance, recall Sergei Eisenstein's account of filmic montage as a technique where individual shots are 'spliced together [. . .] with no transitional material, creating disjunctions in the finished text' to reveal 'a third something' produced only by juxtaposition.[73] Yet artistic intent remains veiled or opaque in Cole's intermedial forms. Expression proceeds by analogy, juxtaposition, and suggestion without the emergence of a coherent or portable message that can be assembled from the fragments.

These tensions never resolve. In keeping with Carter Mathes's analysis of sound in African American literature, inter-artistic experimentation results in uncertainty and complication: 'sonic effects and states such as dissonance, vibration, and resonance [. . .] bridge the aural, the literary, and the political, complicating ideas and representations of time, memory, and narrative perspective'.[74] For Cole, such complexity is part of a broader project of unsettling or troubling readers by complicating any straightforward understanding of art and the world. In an interview with Max Liu in 2011, he explains that he's 'attracted, in art, to things that trouble the complacency of the viewer or reader' and that 'there's no such thing as a right to remain untroubled'.[75] Unease and worry are dominant notes in his work: 'My ambition is – with my photography and with my writing – that there will always be something wrong'.[76]

Take, as an example, Cole's set-piece description of Julius's visit to Central Park for a picnic with friends, where they are 'part of a crowd of city dwellers in a carefully orchestrated fantasy of country life'.[77] With its layering of vivid vignettes, sounds, and silhouettes in the foreground and background – from a toddler named Anna being chased by her father and 'purple blooms' on 'a wisteria's bough hung low' to a carefully choreographed protest staged by parachutists whose fulfilment of 'the ancient dream of flight unfolded in silence' triggers Julius's childhood memory of rescuing a boy who was drowning in a Lagos swimming pool – the scene draws inspiration from Pieter Brueghel the Elder's *Landscape with the Death of Icarus*, a painting that sidelines its apparent subject.[78] Taken together with Julius's reflection upon W. H. Auden's 'Musée des Beaux Arts' (a poem inspired by Brueghel's painting) when he visits Brussels, the allusion hints at Julius's marginalisation of certain narratives and experiences in his own narration.

Cole's creation of a tableau or frieze that calls for painstaking interpretation of the complex interplay between allusions and fragments is not simply a postmodernist trick to 'keep the book heterogeneous'; navigating tensions, evasion, and silences is fundamental to any appreciation of the ethics of the novel.[79] On second reading, it is difficult to miss the disjunction between the heroic pretensions that animate Julius's imaginings of his childhood in Africa and the troubling implications of his objectification of Moji, whom he dismisses as 'too tall' and 'not beautiful in the way [he] expected dark women to be' even as he lingers on the 'sight of the dark curve of her breast'.[80] Armed with the knowledge of Julius's rape of Moji when he was a schoolboy in Nigeria, readers are invited by these fleeting glimpses of Moji's body in pieces to reflect upon the extent of their complicity in overlooking sexual violence that is hiding in plain sight.

Critical Listening

The connections between Cole's formal choices, especially his scrambling of sound, word, and image, and the demand for a mode of critical attention that moves across genres, styles, and the senses come into sharp relief if we explore audio-visual tensions in Cole's (creative and critical) output as a writer and photographer since the publication of *Open City*. In 'The White Savior Industrial Complex' (2012), one of Cole's best-known essays, he criticises the Pulitzer Prize–winning journalist Nicholas Kristof for failing to 'think constellationally. He does not connect the dots or see the patterns of power behind isolated "disasters"'.[81] The initial impetus for Cole's indictment of racialised 'American sentimentality' was entrenched white US rhetoric about Africa that falls back on simplistic narratives about white North American 'saviors' and Black African 'victims', but a demand for nuanced understanding that eschews binary thinking and looks beyond the surface underpins his oeuvre more generally. The audio-visual tensions in *Open City* create spaces for readers to 'think constellationally' and to identify underlying power dynamics. In Cole's subsequent work across various media – especially the phototext and the critical essay – this quest for forms that demand active, if uneasy engagement with complexity, doubt, and confusion has intensified.

In critical essays, both those collected in *Known and Strange Things* (2016) and those printed in his 'On Photography' column (2015–19), Cole has elaborated his thinking about the cultural and political import of noise and quiet and the dialogue between them. He deploys the term 'visual noise' to describe the deleterious effect of digital media's infiltration of our lives and culture. Amid what he calls the 'swirl of visual noise', Cole maintains faith in the expressive power of photography to reduce the volume in a manner that is painful but clarifying. In 2013, he likened a good photograph to a 'pinprick' that 'draws blood', making viewers 'uncomfortable' by delivering a 'momentary shock' that forces them to pay attention. Photography, he suggests, carries the potential to affect the mind and body in a visceral, even violent way (it delivers 'a sudden rush of blood to the head').[82]

A more recent essay recapitulates the trope of 'a world of deafening images', celebrating the 'quiet consolations of photobooks'.[83] Arguing that the 'labor-intensive and fiscally-unsound character' of photobooks marks them as a 'quixotic affront' to the economic logic governing the cultural marketplace, he locates the genre in a realm of spirituality and quietness that leaves readers 'mesmerized by the silent and unanticipated intensity'.[84] Phototexts foster imaginative freedom and daydreaming ('wander[ing] off the beaten path') that is at odds with the automatic reflex of scrolling through pictures on a digital device: 'Sitting with it, you have to sit with yourself: this is a private experience in a time when those are becoming alarmingly rare, an act of analog rebellion in an obnoxiously digital world'.[85] Just as in *Open City*, stillness, slowness, and quietness are aligned with imaginative engagement and thoughtful reflection, but interference and dissonance hover at the edges of any transcendent experience.

Allied to this critique of the capacity of 'visual noise' to blunt critical and creative sensibilities is Cole's indictment of 'the relentless, deadening noise of populism and demagoguery'.[86] Speaking about *Blind Spot* (2017), a phototext assembled against the backdrop of the rise of Donald Trump, Cole explains that his longstanding pursuit of density and opacity became more urgent in a period of political turmoil. *Blind Spot* was conceived as a counterpoint to a 'reductive, simplistic, unanalytical view of the world. This is a time for protest and activism for sure, but it is also a time for subtlety, ambiguity and complexity'.[87] Notwithstanding his paean to the contemplative qualities of the phototext, which he contrasts with the 'unvocalised chatter' that is so fundamental to the experience of reading a novel, Cole's phototexts rarely retreat into an idealised realm of quiet and consolation because they are animated by a fascination with interruption, dissonance, and complexity.[88]

In *Blind Spot*, the centrality of critical listening to Cole's ethical and aesthetic project is evident in his experimentation with the generic conventions of the phototext. Audio-visual tensions are accentuated in order to focus attention on the complex interplay between sound, silence, and the visual image. Cole's most striking innovation concerns the abandonment of captions in favour of lyrical fragments of prose that he calls 'voice-overs'. Rather than appearing as a label or description under the photographs, the 'voice-over' occupies the left-hand side of a two-page spread (the untitled image is printed on the right), an important component of 'a lyric essay that combines photography and text'.[89] Cole's 'voice-overs' are positioned under titles that list the place where the photograph was taken (Beirut, Brazzaville, Black River, Bombay, McMinnville, Wannsee, and so on), but the prose that follows is poetic, philosophical, and reflective. Some fragments take the photograph as the point of departure for digressive meditations on such varied topics as memory, poetry, Russian film, disaster, and visual culture, but others engage with the image indirectly. Cole's aesthetic of assemblage assigns readers an active role: they must identify sources, seek out knowledge, and identify connections to piece together the implications of his opaque, oblique form. Specific images and motifs – including drapes, infrastructure, signs, vision, mountains, ladders, tombs, waking, memory, and resurrection, to mention only a few – are repeated (often with subtle variation), creating what Cole calls 'the singing line that connects them all'.[90] The lyrical 'voice-overs' are also dense with allusions and quotations, which are sometimes integrated into the text in such a way as to frustrate identification. Consequently, the imaginative and ethical possibilities of making connections (both within and beyond the phototext) become cardinal principles of the volume.

A fascination with how images and words can resonate or 'sound' is an important dimension of the 'voice-over' as an intermedial form. Take, for instance, a photograph of Brienzersee in Switzerland that spotlights the textures and colours of a ship's foghorn – freshly painted in white with a gold rim – gleaming against a grey lake, with clouds and mountains in the background. The voice-over is lyrical but compact: 'I opened my eyes. What lay before me looked like the sound of the alphorn at the beginning of the final movement of Brahms's First Symphony. This was the sound, this was the sound I saw'.[91] The words, which reference a fanfare, introduce dynamism and flow to the still image, transforming the foghorn into a megaphone for orchestral music inspired by alpine instruments. Yet the interplay between sound and image is also tinged with loss because the music never sounds. Just as in the descriptions of Brewster's paintings in *Open City*, sound can only be imagined or dreamed (and even that is contingent upon a reader's knowledge of Brahms's symphony). Moreover, the repetition and assonance in the fragment creates complex, intersecting poetic, musical, and emotional effects.

Cole's interest in 'seeing' sound also sheds light on the description of the textual fragments in his phototext as 'voice-overs'. The word, of course, is usually associated with the moving rather than still image, not least the narration that accompanies films and documentaries, which tends to interpret images on screen rather than describing them. In keeping with Cole's layering of lyric forms, the 'voice-over' also invokes the speaking voice, casting the reader as a listener. The idea of a voice reading or speaking as time passes disrupts the stillness of the photograph, introducing tension and interplay between past, present, and future to a form (photography) that Cole regards as 'necessarily an archival art'.[92] Likewise, words and image do not coalesce into transparent meaning; instead, they run in parallel, flowing in conflicting directions that leave room for active interpretation and imaginative engagement. In this respect, it is significant that Cole insists that the 'voice-over' should be understood as a narrative rather than a descriptive mode: it 'has to meet the demands of storytelling, of obliqueness, of compression. It has to detonate in a certain way that might actually be adjacent to the photograph, not sitting right on top of it'.[93] In contrast to T. S. Eliot's famous comparison, in 'Tradition and the Individual Talent' (1919), of the author to a catalyst (necessary to provoke a scientific reaction but inert and aloof), Cole's version of artistic 'detonation' is concerned with neither the creation of new substances nor the author's relationship to the text. Instead, audio-visual tension generates new possibilities for comparison, connection, and disjunction that resist the simplistic consolations of harmony and overlap, looking instead to the productive friction created by an aesthetics of juxtaposition that positions words, sounds, and images 'as associated, as necessarily juxtaposed, if not fully interlinked, or self-consciously interwoven'.[94]

Conclusion

In *Stuart Hall's Voice*, a work of commemoration conceived as a 'long, imaginary, epistolary dialogue' with his late friend and mentor, David Scott argues that speaking and listening rather than critique underpinned Hall's practice as an intellectual and a speaker. Listening and learning, grounded as they are in reciprocity, generosity, and openness, guided Hall's formulation of 'an ethics [and aesthetics] of *responsiveness* to difference'.[95] Scott's attention to the critical and ethical potential of listening is noteworthy because (with some significant exceptions) analysis of 'voice' in Black diasporic literature has tended to emphasise what Henry Louis Gates, Jr. has called 'the speakerly text', a term developed to describe a long tradition of Black writers exploring the creative and political 'possibilities of representation of the speaking black voice in writing'.[96]

The art and ethics of listening also lie at the heart of Teju Cole's creative work across various genres, styles, and media. In interviews and essays, he has often returned to concepts of noise and quietness when considering what role contemporary artists and critics should play in a world defined by the ubiquity of social media and the clamour of populism. Cole's critical and creative engagement with 'the art of listening' rarely strikes the utopian note sounded in Scott's tribute to Hall's generosity. Instead, listening is compromised, damaged by asymmetrical power dynamics, and tarnished by entrenched racism and sexism. Impaired (or inadequate) listening is a recurring motif in Cole's *Open City*, and these failures ricochet back on the reader, raising questions about complicity and agency. Nevertheless, Cole's restless quest for new forms, which combine word, image, and sound in jagged patterns of harmony and disjunction, is motivated by an urgent call to 'think constellationally', listening out for 'the singing line' of 'interconnection' in a world increasingly defined by division, rigid standardization, and unthinking populism.

Notes

1 Erica R. Edwards, 'The New Black Novel and the Long War on Terror', *American Literary History*, 29 (2017), 664–81 (669).

2 Werner Sollors, 'Cosmopolitan Curiosity in an Open City: Notes on Reading Teju Cole by Way of Kwame Anthony Appiah', *New Literary History*, 49 (2018), 227–48 (239). See also Madhu Krishnan, 'Postcoloniality, Spatiality and Cosmopolitanism in the *Open City*', *Textual Practice*, 29 (2015), 675–96; and Pieter Vermeulen, 'Flights of Memory: Teju Cole's *Open City* and the Limits of Aesthetic Cosmopolitanism', *Journal of Modern Literature*, 37 (2013), 40–57.

3 Aaron Bady, 'Interview: Teju Cole', *Post45* (19 January 2015): <https://post45.org/2015/01/interview-teju-cole/> [accessed 20 January 2021].

4 See Alexandra Kingston-Reese's 'Slant Rhyming' for an illuminating discussion of a subset of contemporary US novelists including Cole and Siri Hustvedt

whose parallel, intersecting careers as art critics and novelists generate 'a practice-based reflexive relation between creative and critical aesthetics' (Alexandra Kingston-Reese, 'Slant Rhyming', *ASAP Journal* (9 July 2018), [n.p.]: <http://asapjournal.com/slant-rhyming-alexandra-kingston-reese/>).

5 See, for example: Karen Jacobs, 'Teju Cole's Photographic Afterimages', *Image & Narrative*, 15 (2014), 87–105; Alexandra Reese-Kingston, *Contemporary Novelists and the Aesthetics of Twenty-First Century American Life* (Iowa: University of Iowa Press, 2019); Sam Reese and Alexandra Kingston-Reese, 'Teju Cole and Ralph Ellison's Aesthetics of Invisibility', *Mosaic: An Interdisciplinary Critical Journal*, 50 (2017), 103–19.

6 Kingston-Reese, *Contemporary Novelists*, p. 61.

7 Teju Cole, 'Pitch Forward' (interview with Teju Cole), *Guernica* (15 March 2013): <www.guernicamag.com/pitch-forward/>

8 Yogita Goyal, *Runaway Genres: The Global Afterlives of Slavery* (New York: New York University Press, 2019), p. 173.

9 Ibid., p. 172.

10 Ibid., p. 209.

11 Rachel Sykes, *The Quiet Contemporary American Novel* (Manchester: Manchester University Press, 2017); and Josh Epstein, '*Open City*'s "Abschied": Teju Cole, Gustav Mahler, and Elliptical Cosmopolitanism', *Studies in the Novel*, 51 (2019), 412–32; Teju Cole, *Open City* (New York: Random House, 2011), p. 9.

12 Cole, *Open City*, pp. 75, 38.

13 Ibid., p. 18.

14 Teju Cole, 'In Praise of the Photobook', *Mack* (27 March 2020): <https://mackbooks.co.uk/blogs/news/in-praise-of-the-photobook-by-teju-cole>

15 Jonathan Sterne, *The Audible Past: Cultural Origins of Sound Reproduction* (Durham: Duke University Press, 2003), pp. 16, 15, 348.

16 Teju Cole, 'The Superhero Photographs of the Black Lives Matter Movement', *The New York Times Magazine* (26 July 2016), [n.p.].

17 Teju Cole, 'The White Savior Industrial Complex', in *Known and Strange Things: Essays* (New York: Random House, 2016), pp. 340–9 (344).

18 Cole, *Open City*, p. 9.

19 Ibid., p. 170.

20 Ibid., p. 171.

21 Ibid., p. 14.

22 Ibid., p. 11.

23 Ibid., p. 9.

24 Ibid.

25 Ibid., p. 238.

26 Ibid., pp. 201, 208.

27 Peter Gay, *Freud: A Life for Our Time* (New York: Norton, 1988), p. 97; Cole, *Open City*, p. 206.

28 Bady, 'Interview: Teju Cole'.

29 Cole, *Open City*, p. 5.

30 Ibid., p. 80.

31 Ibid.

32 Jennifer Stoever, *The Sonic Color Line: Race and the Cultural Politics of Listening* (New York: New York University Press, 2016), p. 232.
33 Cole, *Open City*, p. 64.
34 Ibid., p. 70.
35 Ibid., p. 171.
36 Ibid., pp. 23, 24.
37 Cole, 'Pitch Forward'.
38 Sykes, p. 176.
39 Cole, *Open City*, p. 209.
40 Ibid., p. 244.
41 Ibid., p. 245.
42 Rebecca Clark, '"Visible Only in Speech": Peripatetic Parasitism, or, Becoming Bedbugs in *Open City*', *Narrative*, 26 (2018), 181–200 (197).
43 Anthony Byrt, 'Interview: Teju Cole' (8 June 2012): <www.noted.co.nz/archive/archive-listener-nz-2012/interview-teju-cole>
44 Teju Cole, *Open City* (excerpt): <https://genius.com/Teju-cole-open-city-excerpt-annotated>
45 Elza Adamowicz, *Surrealist Collage in Text and Image: Dissecting the Exquisite Corpse* (Cambridge: Cambridge University Press, 1998), p. 25.
46 Sean O'Hagan, 'Interview: Teju Cole: My Camera Is Like an Invisibility Cloak', *The Guardian* (25 June 2017): <www.theguardian.com/books/2017/jun/25/teju-cole-blind-spot-my-camera-is-like-an-invisibility-cloak-interview>
47 'Page for and about *Open City*': <https://op-cit.tumblr.com/>
48 Ibid.
49 Steve Paulson, 'Finding My Way into a New Form: An Interview with Teju Cole', *The Millions* (5 July 2017): <https://themillions.com/2017/07/finding-way-new-form-interview-teju-cole.html>; James M. Hoddap, 'Lumpy Sympathies: An Interview in Beirut with Teju Cole', *Research in African Literatures*, 48 (2017), 243–9 (246).
50 Michael Baxandall, *Patterns of Intention: On the Historical Explanation of Pictures* (New Haven: Yale University Press, 1985), p. 5; Anderson Tepper, 'A Conversation with Teju Cole', *Tin House* (26 January 2011): <https://tinhouse.com/a-conversation-with-teju-cole/>
51 Tepper, 'A Conversation with Teju Cole'.
52 Cole, *Open City*, p. 37.
53 Ibid., pp. 38–9.
54 Ibid., pp. 37–8.
55 Ibid., p. 38.
56 Ibid., pp. 38, 39.
57 Ibid., p. 38.
58 Ibid., pp. 37, 40.
59 Ibid., p. 138.
60 Ibid., p. 137.
61 Ibid., p. 138.
62 Ibid.
63 Ibid.
64 Ibid., p. 6.
65 Ibid., p. 140.

66 Ibid., p. 138.
67 Ibid., p. 140.
68 Teju Cole, 'A True Picture of Black Skin', in *Known and Strange Things*, pp. 144–51 (150, 151).
69 Ibid., p. 148.
70 Cole, *Open City*, p. 138.
71 King-Kok Cheung, *Articulate Silences: Hisaye Yamamoto, Maxine Hong Kingston, and Joy Kogewa* (Ithaca: Cornell University Press, 1993).
72 Cole, *Open City*, p. 138.
73 Anne Elizabeth Carroll, *Word, Image, and the New Negro: Representation and Identity in the Harlem Renaissance* (Bloomington: Indiana University Press, 2005), pp. 27–8.
74 Carter Mathes, *Imagine the Sound: Experimental African American Literature after Civil Rights* (Minneapolis: University of Minnesota Press, 2015), p. 6.
75 Teju Cole interviewed by Max Liu, 'Palimpsest City', *3: AM Magazine* (16 August 2011): <www.3ammagazine.com/3am/palimpsest-city/>
76 Cole, 'Pitch Forward'.
77 Cole, *Open City*, p. 194.
78 Ibid., pp. 197, 195.
79 Tepper, 'A Conversation with Teju Cole'.
80 Cole, *Open City*, pp. 198, 199.
81 Cole, 'The White Savior Industrial Complex', in *Known and Strange Things*, pp. 340–9 (344).
82 Kishwer Vikaas, 'Teju Cole: A Good Photograph Is Like a Pinprick', *The Aerogram* (11 March 2013): <http://theaerogram.com/teju-cole-on-photography/>
83 Cole, 'In Praise of the Photobook'.
84 Ibid.
85 Ibid.
86 O'Hagan, 'Interview with Teju Cole'.
87 Ibid.
88 Cole, 'In Praise of the Photobook'.
89 Teju Cole, *Blind Spot* (London: Faber, 2017), p. 324.
90 Ibid.
91 Ibid., p. 120.
92 Teju Cole, 'Disappearing Shanghai', in *Known and Strange Things*, pp. 165–70 (165).
93 Paulson, 'Finding My Way into a New Form'.
94 Susan Gillman and Alys Eve Weinbaum, *Next to the Color Line: Gender, Sexuality, and W. E. B. Du Bois* (Minneapolis: University of Minnesota Press, 2007), p. 3.
95 David Scott, *Stuart Hall's Voice: Intimations of an Ethics of Receptive Generosity* (Durham: Duke University Press, 2017), p. 115.
96 Henry Louis Gates, Jr., *The Signifying Monkey: A Theory of African-American Literary Criticism* (Oxford: Oxford University Press, 1988), p. xxv.

Earpiece 3

'Really a Part of Me'

Dementia Conversations

Catherine Charlwood

More than anything, this is about listening. Listening to what's said rather than to what you hear within it. It's also about dementia, my grandma, music, and the jazz standards which emanated from Tin Pan Alley. Take your pick. Part of this is an unashamed reflection on my family's musical preferences, but this leads me to larger questions about the nature of identity, memory, and the power of narrative.

My grandma, Doreen, was born in 1925, at the height of the Charleston craze. She loved singing and dancing: she remembered the war with great fondness for 'all the dances with men in uniform – and then the Americans came!' with new uniforms and new dance steps. As children, my sister and I both learned music and went to tap dancing lessons. Sinatra was a staple, as were several swing and jazz standards which our generation are simply not meant to know. The music of Gershwin, Kern, Porter, and Berlin, then, was a shared interest between us and our grandparents from the off.

'With feeling' is a direction sometimes written at the top of a piece of music. Indeed, Gershwin's 'They Can't Take That Away From Me' – of which more later – is to be sung 'lightly, and with feeling'.[1] It is this addition of 'feeling', so intrinsic to music and memory, on which I wish to focus, in two distinct ways. One is the emotion that music engenders, both at the time of its playing, and for years afterwards in memory. The other is music's capacity to afford one physical feelings: its ability to be embodied in this case, through dance.

The lyrics of many of the songs I'm thinking of are fanciful, it's true. Sinatra can loop the melody upwards as he sings 'and if you should survive, to 105, / look at all you'll derive / out of being alive!', but that is something which centenarians may or may not feel in reality.[2] As frailty crept in, Grandma acquired a stick and then a tri-walker. In motion, she was a dab hand (should that be 'foot'?), but it was beginning that was the difficulty (cue the Billie Holiday recording of 'I Can't

DOI: 10.4324/9780429291616-13

Get Started'). She would make several false starts at walking, nudging repeatedly at the ground but not transferring her weight. However, as soon as we sang to her, got her humming or singing, she would walk without a problem in time to the music, as if unaware there had ever been a hesitation: 'I've told ev'ry little star, / Just how sweet I think you are' and she was off, in effect dancing in walk format.[3] As Oliver Sacks notes, people can 'regain a fluent motor flow with music', and we witnessed this transformation: put on the music, and the dancer returns.[4] Indeed, 'music [. . .] can be invoked as an ally for a variety of world-making activities', and this was certainly true for Grandma, because it got her from room to room, to a meal, in or out of the car, and, beyond the instrumental (pun very much intended), reconnected her with an important aspect of her identity.[5]

As Tia DeNora notes, 'music is an active ingredient in the organization of self' and does not

> simply *act upon* individuals, like a stimulus. Rather, music's 'effects' come from the ways in which individuals orient to it, how they interpret it and how they place it within their personal musical maps, within the semiotic web of music and extra-musical associations.[6]

All of the different associations which connected Grandma to music helped preserve certain memories. Music served a social function (she readily remembered her father playing the piano at family parties 'though he never took a lesson in his life') and was often indivisible from movement, given that she line-danced into her eighties. My own dancing and my sister's also seemed to spark some jollity of recognition of a meaningful activity even if she could no longer join in herself (or, sometimes, necessarily tell us who we were).

The lyrics of Cole Porter's 'I've Got You Under My Skin' kept coming back to me as Alzheimer's played its own tune, especially 'I've got you under my skin, / I've got you deep in the heart of me, / So deep in my heart, that you're really a part of me. / I've got you under my skin.'[7] Allen Forte refers to Porter's 'witty and poignant lyrics', and the latter was certainly true for me.[8] I began to worry about the nature of that 'really'. What does it mean to be 'really a part of me' when the 'me' implicated is losing parts of her memory? Could she remain 'really a part of me' and I of her? In attributing certain songs to Grandma, I was insisting on these as a part of her individual makeup despite knowing that their prominence in her life (and memory) was in part a by-product of their being so often played on the radio and in the dance hall (in her youth) and ubiquitously in shopping centres and TV adverts now.[9] However, for all that being popular might seemingly make these songs less personal, 'in songs, words are the sign of a voice'. As Simon Frith continues:

> Songs are more like plays than poems; song words work as speech and speech acts, bearing meaning not just semantically, but also as structures of sound that are direct signs of emotion and marks of character.[10]

'Song words', lyrics, can also inspire what is not there: the music. Using my now-ailing iPad, I recorded Grandma singing with me and Mum in 2015. First, we gave her the first words to 'I've Told Ev'ry Little Star'; using the rhyme and metre, she could continue. When Mum gave her 'the words' to 'A Nice Cup of Tea' – one piece of the jigsaw – Grandma could supply several other puzzle pieces: the notes, the rhythm, the cadences that come of making meaning from all of these elements.[11] Those recordings, nonchalant as they were at the time, are such a source of pleasure now. It is her distinctive voice, giggle, turn of phrase, and her instinctive musical leanings (regardless of what she understood at the time or would remember later). When I hear her again, she is here.

The other song which inevitably worried me was 'They can't take that away from me', because dementia was demonstrably capable of a certain amount of theft. Gershwin's deft lyrics focus in on those idiosyncrasies so indicative of individual personality:

The way you wear your hat,
The way you sip your tea,
The mem'ry of all that –
No, no! They can't take that away from me![12]

It is a song which, perhaps, protests too much: there are lots of double negatives ('no, no!'; 'never, never'). But the real focus of the lyrics 'is the utterly simple *that* [. . .]. On the one hand, "that" seems to minimize the loss of "all *that*" yet at the same time to cherish the indefinable essence of character captured in the unbegrudging common place catch-phrase'.[13] Dementia can, of course, also change behaviours in the present, regardless of memory: the way 'you sip your tea' or 'hold your knife' may change through dementia (though it may be more palatable to accept that 'the way we danced till three' has now gone).[14] While Gershwin gives specific examples, though, it is 'the insignificant linguistic bit "that"' which takes centre stage, where 'in this context, it hints at unspoken depths of feeling'.[15] If we unmoor from specifics to the more generic 'that', then it was not taken away from either her or me. There was always something indefinitely Grandma: you might not be able to put your finger on it, but that was it, and it was that.

In insisting that lyrics are freighted with meaning, though, I was refusing to hear music holistically, since '[a]s a neurobiological phenomenon, [music] is multidimensional: these dimensions range from the decoding of abstract sensory signals potentially lasting several hours, to physiological responses'.[16] I was also being selfish, making it more about me than anyone else: noticeably, both of the songs which played on my mind revolve around the pronoun 'me', a subject position which either my grandma or I could occupy. Although I was listening, I was over-hearing things in the lyrics. The literary critic in me was falling into the pitfalls of an over-easy reliance on content analysis, harking back to 'the sociology of popular music' in the 1950s and 1960s, when it was 'dominated by analyses of song words'.[17] I was worrying about the lyrics when I could, like Grandma, have been carried by the song.

While I trained as a literary scholar, I now work in research translation and impact within dementia research, a role which offered an unusual alignment of my personal and professional interests. I work for the IDEAL programme at the University of Exeter, which studies the factors which make a difference to living a good quality of life with mild to moderate dementia, in the case of people with dementia and of their carers living in the community.[18] One of the best parts of my role is enabling the voices of people with dementia and of carers to be heard. We co-created the Living with Dementia Toolkit, and ever since its launch the most-commented-on feature are the videos: hearing directly from those experiencing dementia.[19] As George, who lives with dementia, says, 'Telling your story is really important. It's your narrative, your story, that will make people listen and change their minds'.[20] When people 'really' listen, their perspectives can shift, even slightly, and that is something very much needed in the world of dementia. The 2019 World Alzheimer's Report on 'Attitudes to Dementia' revealed that 78% of the global survey's 70,000 respondents were concerned about developing dementia in the future.[21] People hear 'dementia' and think only the worst, but while dementia obviously brings challenges, many people live interesting, fulfilling lives for many years after diagnosis. When you speak to people with dementia and carers, what they want is hope: to hear something different from the largely negative narrative in the press and, too often, in healthcare appointments.

Listening matters. It matters because only by listening to those living with a condition or to those caring for those living with a condition will we really learn what makes a difference to those lives. There are also practical reasons to pay attention to listening and dementia. Hearing loss is a known risk factor for developing dementia, and for those who have dementia, hyperacusis (noise sensitivity) can be a pervasive symptom.[22] During IDEAL's COVID-19 research, we heard from many people who were nervous about emerging into the aural environment of a post-lockdown world, a world where

> [t]here were too many people in the restaurant and I could hear them talking and I was getting distracted and couldn't understand what the people I was with were saying. I used to try and make an excuse not to go inside buildings – I only recognised this as an issue in August [2020] when the restrictions eased.[23]

Indeed, we shared on the toolkit a co-produced resource about coping with this specific aspect of getting back out into the world.[24]

In another video on the Living with Dementia Toolkit, I interview Angela, who used to care for her Jamaican mother who had dementia. 'She had a great love of film musicals', Angela remembers, 'although she didn't recognise various people'; *The Sound of Music* came on by chance one day, during a moment of distress, 'and she started singing along to it. And she knew all the words!'[25] Musical memories and their meanings are cross-cultural, too. My perspective on Porter's lyrics has shifted with time and experience. Originally, I heard 'really' as an intensifier, an assurance of veracity: 'you're *really* a part of me [in an irrefutable, provable way, and] I've got you under my skin'. Now, I hear it as a comfort: in the version of my grandma who has as much agency and free will as she did pre-Alzheimer's, all of her memories *are* there. This is 'really' as qualifier rather than intensifier: really, they were a part of her, even when her current self could not access them.[26]

Love was a feeling that could be abstracted from my grandma's ability to negotiate (in any self-willed way) the content of her memories: she could still feel it and did.[27] The love of music demonstrably never left her. I cannot say that I would not be where I am today if it were not for my grandma and our own experience of dementia. I still hear her voice ('goody gumdrops!') and the songs and melodies which became so familiar. What always was really a part of her has become really a part of me, and for that, I'm grateful. And I'll keep on listening.

Notes

1 Ira Gershwin, *Lyrics on Several Occasions* (London: Elm Tree Books, Hamish Hamilton, 1977), p. 47.
2 Johnny Richards and Carolyn Leigh, *Young at Heart*, Frank Sinatra, Nelson Riddle, Nelson Riddle Orchestra (Sunbeam Music, 1953).
3 Oscar Hammerstein II, 'I've Told Ev'ry Little Star', in *Lyrics* (Milwaukee, WI: Hal Leonard Books, 1985), p. 68. The music was composed by Jerome Kern.
4 Sacks is here writing about Parkinson's disease, but the point stands (Oliver Sacks, *Musicophilia: Tales of Music and the Brain* (New York; Toronto: Alfred A. Knopf, 2008), p. 345).

5 Tia DeNora, *Music in Everyday Life* (Cambridge: Cambridge University Press, 2000), p. 40.
6 Ibid., p. 61 (original emphasis).
7 Cole Porter, 'I've Got You under My Skin', in *The Complete Lyrics of Cole Porter*, ed. Robert Kimball (Cambridge, MA: Da Capo Press, 1992), p. 201.
8 Allen Forte, 'Secrets of the Melody: Line and Design in the Songs of Cole Porter', *The Musical Quarterly*, 77 (1993), 607–47 (607).
9 My examples here are in part drawn from Geoffrey Block, who observes of musicals songs from the 1920s and 1930s that 'even when heard in instrumental versions in shopping centres, lifts and cafes, many retain at least a recognisable association with their original verbal messages' ('The Melody (and the Words) Linger on: American Musical Comedies of the 1920s and 1930s', in *The Cambridge Companion to the Musical*, ed. William A. Everett and Paul R. Laird, 2nd edition (Cambridge: Cambridge University Press, 2008), pp. 103–23 (p. 112)). Philip Furia notes Ira Gershwin's dislike of the overplaying of 'Love Walked In': 'Overexposure by radio play, he foresaw, could quickly exhaust a song's endurance, leading to increasingly rapid turnover of songs and a falling off in craftsmanship' (*Ira Gershwin: The Art of the Lyricist* (New York; Oxford: Oxford University Press, 1994), p. 155).
10 Simon Frith, 'Why Do Songs Have Words?', *Contemporary Music Review*, 5 (1989), 77–96 (90).
11 I did not know it at the time, but this is similar to one of the tests Cuddy et al. conducted: 'Lyrics prompt: given the lyrics of the first phrase of a familiar song spoken in a monotone, participants are asked to sing a continuation of the song' (Lola L. Cuddy, Ritu Sikka and Ashley Vanstone, 'Preservation of Musical Memory and Engagement in Healthy Aging and Alzheimer's Disease' (*The Neurosciences and Music*, V), *Annals of the NY Academy of Sciences*, 1337 (2015), 223–31 (225)).
12 Gershwin, *Lyrics on Several Occasions*, p. 47.
13 Furia, *Ira Gershwin*, pp. 142–3.
14 Gershwin, *Lyrics on Several Occasions*, pp. 47–8.
15 Philip Furia, *The Poets of Tin Pan Alley: A History of America's Great Lyricists* (New York; Oxford: Oxford University Press, 1992 [1990]), p. 146.
16 Camilla N. Clark and Jason D. Warren, 'Music, Memory and Mechanisms in Alzheimer's Disease', *Brain*, 138 (2015), 2122–5 (2125).
17 Frith, 'Why Do Songs Have Words?', p. 77.
18 IDEAL Programme: <http://idealproject.org.uk> [accessed 5 October 2022]. IDEAL was funded by NIHR and ESRC 2014–2018 and by Alzheimer's Society as a Centre of Excellence since 2018.
19 IDEAL Programme, Living with Dementia Toolkit: <http://livingwithdementiatoolkit.org.uk> [accessed 5 October 2022].
20 IDEAL Programme, 'Why Is It Important to Talk about Dementia?' (online video recording), in 'Talking Honestly about Dementia', Living with Dementia Toolkit: <http://livingwithdementiatoolkit.org.uk/stay-positive/talking-honestly-about-dementia> [accessed 10 October 2022].
21 Alzheimer's Disease International, *World Alzheimer Report 2019: Attitudes to Dementia* (London: Alzheimer's Disease International, 2019), p. 14.

22 See Rhett S. Thomson, Priscilla Auduong, Alexander T. Miller and Richard K. Gurgel, 'Hearing Loss as a Risk Factor for Dementia: A Systematic Review', *Laryngoscope Investigative Otolaryngology*, 2 (2017), 69–79.

23 IDEAL Programme, 'Tips for Coming Out of Lockdown', Living with Dementia Toolkit: <http://livingwithdementiatoolkit.org.uk/stay-safe-and-well/tips-for-coming-out-of-lockdown> [accessed 10 October 2022].

24 DEEP: The UK Network of Dementia Voices, 'Coming Out of Lockdown – Coping with Noise Sensitivity': <www.dementiavoices.org.uk/coming-out-of-lockdown-coping-with-noise-sensitivity> [accessed 10 October 2022].

25 IDEAL Programme, 'Angela Shares Her Dementia Story' (online video recording), in 'Connecting with Your Culture', Living with Dementia Toolkit: <http://livingwithdementiatoolkit.org.uk/stay-connected/connecting-with-your-culture> [accessed 10 October 2022].

26 This sentence is hard to write, since I first penned it in the present tense. Doreen died in 2020.

27 In IDEAL's recent documentary film *The World Turned Upside Down* (2022), I share the story of how my grandma could not recognise my uncle, who was living with her, 'but I know I love him'. This felt, though not understood, love was what kept her happy to be around him.

Index

9780367261306